Single-Arm Phase II
Survival Trial Design

Chapman & Hall/CRC Biostatistics Series

Series Editors

Shein-Chung Chow, Duke University School of Medicine, USA
Byron Jones, Novartis Pharma AG, Switzerland
Jen-pei Liu, National Taiwan University, Taiwan
Karl E. Peace, Georgia Southern University, USA
Bruce W. Turnbull, Cornell University, USA

Recently Published Titles

Interface between Regulation and Statistics in Drug Development
Demissie Alemayehu, Birol Emir, Michael Gaffney

Innovative Methods for Rare Disease Drug Development
Shein-Chung Chow

Medical Risk Prediction Models: With Ties to Machine Learning
Thomas A Gerds, Michael W. Kattan

Real-World Evidence in Drug Development and Evaluation
Harry Yang, Binbing Yu

Cure Models: Methods, Applications, and Implementation
Yingwei Peng, Binbing Yu

Bayesian Analysis of Infectious Diseases
COVID-19 and Beyond
Lyle D. Broemeling

Statistical Meta-Analysis using R and Stata, Second Edition
Ding-Geng (Din) Chen and Karl E. Peace

Advanced Survival Models
Catherine Legrand

Structural Equation Modeling for Health and Medicine
Douglas Gunzler, Adam Perzynski and Adam C. Carle

Signal Detection for Medical Scientists
Likelihood Ratio Test-based Methodology
Ram Tiwari, Jyoti Zalkikar, and Lan Huang

Single-Arm Phase II Survival Trial Design
Jianrong Wu

For more information about this series, please visit: https://www.routledge.com/
Chapman--Hall-CRC-Biostatistics-Series/book-series/CHBIOSTATIS

Single-Arm Phase II Survival Trial Design

Jianrong Wu

CRC Press
Taylor & Francis Group
Boca Raton London New York

CRC Press is an imprint of the
Taylor & Francis Group, an **informa** business

A CHAPMAN & HALL BOOK

First edition published 2022
by CRC Press
6000 Broken Sound Parkway NW, Suite 300, Boca Raton, FL 33487-2742

and by CRC Press
2 Park Square, Milton Park, Abingdon, Oxon, OX14 4RN

© 2022 Jianrong Wu

CRC Press is an imprint of Taylor & Francis Group, LLC

ISBN: 9780367653453 (hbk)
ISBN: 9780367653491 (pbk)
ISBN: 9781003129059 (ebk)

Typeset in CMR10 font
by KnowledgeWorks Global Ltd.

Contents

Preface

The early phase clinical trial has become a key component for successfully developing advanced cancer treatments. The single-arm phase II trial is an initial step to access the primary efficacy of a new agent and drug. Traditional cancer treatments are chemotherapy or radiation or chemotherapy plus radiation. The primary endpoint of cancer clinical trials is tumor response which is often categorized as responder (complete response or partial response) or non-responder (stable disease or disease progression). However, chemotherapy or radiation has not been successful for treating patients with advanced cancers.

Recently, advances in the understanding of tumor biology, success in sequencing of cancer genomes and increasing knowledge of the molecular basis of anti-tumor immune response have led to an increase in molecularly targeted, genomic-driven and immunotherapy clinical trials where antitumor activity based on tumor shrinkage cannot be directly hypothesized and more appropriate endpoints are recommended. Specifically, progression-free survival (PFS) or time to disease progression (TTP) have been proposed. Furthermore, randomized trials are recommended for these trials due to lack of appropriate historical data and the subjective nature of assessing disease progression. However, a randomized trial needs a lot more patients which is often not achievable for numerous reasons, including logistic feasibility, funding, and time limitations, etc. Therefore, single-arm phase II trials with time-to-event endpoints are often conducted to access the primary efficacy of molecularly targeted therapy or immunotherapy.

This book is intended to provide a comprehensive summary of the most commonly used methods for single-arm phase II trial design with time-to-event endpoints. Some materials in this book are being published here for the first time. I hope this book will serve as a reference book for the researchers conducting clinical trials and for statisticians designing early phase clinical trials for the pharmaceutical industry and cancer research institutes. It may also serve as a textbook for graduate students in the biostatistics field who wish to learn clinical trial design and sample size calculation for early phase clinical trials.

The main focus of this book is on the methodology of single-arm phase II trial design with time-to-event endpoints and illustrated using data from real cancer clinical trials. The book has nine chapters, Chapter 1 introduces concepts of single-arm phase II trial designs and provides an overview of the available literature for such trial design. Chapter 2 introduces single-arm phase II trial design under the Weibull model using log-transformed and quadratic transformed MLE tests. Chapter 3 presents single-arm phase II trial design evaluating survival probability at a landmark time point using Nelson-Aalen estimate based tests and Kaplan-Meier estimate based tests. In Chapter 4, optimal two-stage designs based on the log-log transformed test and the arcsin square root transformed test are given. Chapter 5 discusses single-arm phase II trial designs using one-sample log-rank test and modified one-sample log-rank test. A general hypothesis testing using one-sample log-rank test is also discussed. In Chapter 6, optimal two-stage designs using one-sample log-rank test with restricted follow-up are presented and optimal two-stage design with unrestricted follow-up is also developed. Chapter 7 presents cancer immunotherapy trial designs with random delayed treatment effect and long-term survivors. Chapter 8 discusses single-arm phase II trial design with modulation growth index as the primary endpoint. Chapter 9 presents Bayesian one-stage and two-stage designs.

In this book, we provide a comprehensive summary of the most recent developments in single-arm phase II trial design with time-to-event endpoints. Furthermore, the book also includes some material which has not been published in the literature. For example, a general hypothesis testing using the one-sample log-rank test; optimal two-stage test with unrestricted follow-up using modified one-sample log-rank test; single-arm phase II cancer immunotherapy trial design using change sign weighted one-sample log-rank test; testing quantile of time-to-progression (TTP) ratio, etc.

Each chapter is essentially self-contained. Thus, a reader interested in a particular topic need not to read the entire book but can turn straight to the chapter and use the sample size formulae and R codes therein for the sample size calculations without knowing the details of mathematical derivation. For convenience, all R codes in the book are available for downloading from the CRC Press website at https://www.crcpress.com.

I have much gratitude for many discussions with my colleagues Haitao Pan, Shengping Yang, Liang Zhu, Hui Zhang and Xiaofei Wang. In addition, I would like to thank the Division of Cancer Biostatistics, Biostatistics and Bioinformatics Shared Resource at the Markey Cancer Center University of Kentucky for their support in this undertaking.

1

Introduction of Single-Arm Phase II Trial Design

1.1 Why a Single-Arm Phase II Trial?

Cancer clinical trials investigate the efficacy and toxicity of experimental cancer therapies through four phases of clinical trials, designated phase I, II, III and IV. A phase III cancer clinical trial is usually a large-scale randomized study to compare the efficacy of a new or experimental treatment to that of the best current standard treatment or placebo, while a phase II cancer clinical trial is often conducted as a single-arm study to determine whether a new treatment has sufficient antitumor activities to warrant further investigation in a large-scale randomized phase III trial. However, a single-arm trial compares to a reference group from historical data and doesn't have a concurrent control group. Thus, the results from a single-arm phase II trial could be biased. However, a single-arm trial requires relatively small sample size and all patients are treated with the new treatment. Furthermore for patients with a rare life-threatening disease, the use of a control arm may be unethical or unfeasible. Therefore, single-arm phase II trials are still frequently used in cancer studies to provide some preliminary efficacy and toxicity assessment for new or experimental treatments.

1.2 Primary Endpoint of a Single-Arm Phase II Trial

The antitumor activity for evaluating cytotoxic compounds might be quantified by tumor response which is often categorized to be a binary endpoint as responder if the patient achieved a complete response (CR) or partial response (PR) or non-responder if the patient had stable disease (SD) or progressive disease (PD), where the response evaluation criteria is often based on response evaluation criteria in solid tumor (RECIST). For the trial design with tumor response rate as the primary endpoint, investigators identify the response rate (p_0) (CR or PR) of the standard treatment from historical or literature data as the null hypothesis, e.g., $p_0 = 30\%$. Therefore, for single-arm phase II trial design, p_0 is assumed known and is not subject to variation even though it may be estimated from historical or literature data or a reference level selected by the investigators. This is a fundamental difference between a single-arm phase II trial and a randomized phase II trial.

1.3 Hypothesis of a Single-Arm Phase II Trial

For a single-arm phase II trial with tumor response rate as the primary endpoint, the research hypothesis is if the new or experimental treatment can improve the tumor response rate for a target patient group. Therefore the research hypothesis is typically set to be as a one-sided hypothesis

$$H_0 : p \leq p_0 \quad vs \quad H_1 : p > p_0,$$

where p is the true response rate of the new treatment which is unknown. If the true response rate of the new treatment is less than or equal to p_0, the new treatment is not promising for further investigation and if the true response rate of the new treatment is greater than p_0, the new treatment is promising for further investigation in the future in a phase III trial.

1.4 Test Statistic and Study Design

To calculate the required sample size, investigators have to choose a response rate of the new treatment (p_1), e.g., $p_1 = 50\%$, where $p_1 - p_0 = 20\%$ is a minimum clinical meaningful effect size to detect for the sample size calculation. For the study design, typically, the reference response rate p_0 is chosen to be a value of the probability of tumor response to the standard treatment for the same disease group, and p_1 or effect size $p_1 - p_0$ is selected to identify a minimum clinically meaningful improvement over the historical value. To design the study, we have to select a test statistic to make the inference to the research hypothesis, e.g., reject null hypothesis H_0. A nature test statistic is $\hat{p} = X/n$ which is an estimated probability of response rate for the experimental treatment, where X is the number of subjects who experienced response (CR or PR) and n is the total number of subjects in the study. However the distribution of $\hat{p}$ is often skewed when sample size is small. To make the distribution more normal, we take an arcsin square root transformation to be $\arcsin\sqrt{\hat{p}}$. Using delta method, it can be shown that

$$\text{var}(\arcsin\sqrt{\hat{p}}) \simeq \frac{1}{4n},$$

then, the standardized test statistic Z is given by

$$Z = \sqrt{4n}(\arcsin\sqrt{\hat{p}} - \arcsin\sqrt{p_0}),$$

which is approximately standard normal distributed under the null hypothesis. A large observed value of Z indicates a large treatment effect, therefore, given a type I error rate α, we reject null hypothesis H_0 if $Z > z_{1-\alpha}$, where $z_{1-\alpha} = \Phi^{-1}(1-\alpha)$ and $\Phi(\cdot)$ is the cumulative distribution function (CDF) of standard normal.

To design a study, we need to calculate the sample size (n) or number of subjects required to have sufficient statistical power $(1 - \beta)$ to detect the treatment effect. The statistical power is the probability to reject null hypothesis H_0 when the alternative hypothesis H_1 is true. Under the alternative hypothesis $H_1 : p = p_1$ $(> p_0)$, the test Z is approximately normal with mean $\sqrt{4n}(\arcsin\sqrt{p_1} - \arcsin\sqrt{p_0})$ and unit variance. Thus, the study power $1 - \beta$ satisfies the following equation:

$$\begin{aligned} 1 - \beta &= P(Z > z_{1-\alpha}|H_1) \\ &\simeq \Phi\{\sqrt{4n}(\arcsin\sqrt{p_1} - \arcsin\sqrt{p_0}) - z_{1-\alpha}\}. \end{aligned}$$

By inverting the CDF $\Phi(\cdot)$ from the above equation, we have

$$\sqrt{4n}(\arcsin\sqrt{p_1} - \arcsin\sqrt{p_0}) = z_{1-\alpha} + z_{1-\beta}.$$

Solving n, we obtain the following sample size formula for the Z test

$$n = \frac{(z_{1-\alpha} + z_{1-\beta})^2}{4(\arcsin\sqrt{p_1} - \arcsin\sqrt{p_0})^2}. \tag{1.1}$$

Sample size calculation using formula (1.1) is available from web-based software offered by the Southwest Oncology Group (SWOG) at https://stattools.crab.org/Calculators/oneArmBinomial.html.

However, single-arm phase II trials are often conducted with small sample size and asymptotic normality of the Z test may be inaccurate. Thus, the type I error rate and power may not be maintained well by using the Z test. Simulation study can be conducted to investigate the empirical type I and power of the test Z. An alternative study design can be carried out based on a one-sample exact binomial test given as follows. For a given type I error rate α, a positive integer r exits for a trial with n subjects such that

$$\sum_{k=r}^{n} b(k, p_0; n) \leq \alpha \quad \text{and} \quad \sum_{k=r-1}^{n} b(k, p_0; n) > \alpha, \tag{1.2}$$

where $b(k, p; n)$ is the binomial probability. The r is called critical value. We reject the null hypothesis H_0 if $d \geq r$, where d is total number of responses observed from n subjects. Thus, given the power $1 - \beta$ under the alternative hypothesis $p = p_1$, the sample size required for the study can be obtained by solving for the smallest integer n that satisfies by the following equation:

$$\sum_{k=r}^{n} b(k, p_1; n) \geq 1 - \beta. \tag{1.3}$$

An iteration procedure is used to solve r and n which satisfy the type I error rate and power equations of (1.2) and (1.3). Sample size calculation using one-sample exact binomial test is available from the STPlan software version 4.5 (Brown et al., 2010).

Be aware that different test statistics will result in different sample sizes. Therefore in the study design we have to specify what test statistic is used for the sample size calculation and that test statistic should be used for the final data analysis to reject the null hypothesis or not.

1.5   Choice of Alpha and Beta

The type I error rate α is the probability of rejecting null hypothesis when the null hypothesis is true. That is a false positive rate and type II error rate β is the probability of accepting the null hypothesis when the alternative hypothesis is true, that is a false negative rate ($1-\beta$ is the power). In single-arm phase II oncology trials, a false positive event is considered more serious than a false negative event because for drug development in cancer treatment, a false positive event could be life-threatening while a false negative event may waste some resources but without harm to the patients. Therefore, the typical choice of the type I error rate α is 0.05 or 0.10 and type II error rate is 0.20 (or 80% of power). A relatively large type I error rate and type II error rate could reduce the sample size and therefore save the cost and time for conducting a single-arm phase II trial which provides some preliminary efficacy and toxicity assessment for the experimental treatment.

1.6   Time-to-Event Endpoints

Even through the tumor response rate is the most common primary endpoint in single-arm phase II oncology trials, there are some situations in which the tumor response may not be an appropriate endpoint, such as cytostatic or molecular targeting agents, which can prevent the growth of a tumor but may not kill the tumor cells and lead to tumor shrinkage. In such a case, time-to-event, e.g., disease progression-free survival (PFS) or event-free survival

(EFS), is an alternative endpoint for the trial design. For single-arm trials, PFS is defined as time (date) from study enrollment to time (date) disease progression, death or date of last follow-up, whichever occurs first (here an event is a disease progression or death). The subject's time-to-progression is censored if the subject did not experience the disease progression at the end of the study. Thus, the observed time-to-event data is right censored incomplete data. Ignoring censoring could introduce bias into the study design and data analysis. Survival methodology can be used to analyze the censored time-to-event data (Kalbfleisch and Prentice, 2002).

If the primary endpoint for the trial is the survival probability at a landmark time point, e.g., 1-year PFS recorded as $S(1)$, the research hypothesis is set to be as a one-sided hypothesis

$$H_0 : S(1) \leq S_0(1) \quad vs \quad H_1 : S(1) > S_0(1),$$

where $S_0(1)$ is the 1-year PFS of the standard treatment identified from historical or literature data. Similar, $S_0(1)$ is assumed known and not subject to variation. A naive approach is to treat the 1-year PFS as a binary endpoint. That is, if a subject is disease progression-free within 1 year, it counts as 1 otherwise; it counts as 0. Then, the proportion of subjects who are disease progression-free within 1 year is the estimate of 1-year PFS. Thus, the primary endpoint of 1-year PFS is transformed as a binary endpoint and the study design can be done using the exact binomial test as discussed in the previous section. However the drawback of the naive approach is that censoring due to patients being lost to follow-up can't be counted. Ignoring censoring will result in a biased estimate of the 1-year PFS rate and the study design is also inefficient. Using appropriate methods for designing single-arm phase II trials with various time-to-event endpoints is among the topics of this book.

1.7   Literature Review

Various researchers have proposed methods for sample size and power calculations for randomized phase III trials with time-to-event endpoints (e.g., George and Desu, 1977; Lachin, 1981; Rubenstein et al., 1981; Schoenfeld and Richter, 1982; Schoenfeld, 1983; Lakatos, 1988 and many others). A summary of these methods and some recent developments for randomized phase III survival trial designs can be found in the book "Statistical Methods for Survival Trial Design" by Wu (2018). Randomized phase II trial designs with a prospective control are summarized in the book "Randomized Phase II Cancer Clinical Trials" by Jung (2013). However, the literature for designing single-arm phase II trials with time-to-event endpoints is relatively sparse. For testing the survival probability at a landmark time point, Case and Morgan (2003) and Huang et al. (2010) developed optimal two-stage designs based on the Nelson-Aalen estimate of survival probabilities. Huang and Thomas (2014) extended the results to three-stage design and implemented in R package 'Optim2Design'. Whitehead (2014) proposed a two-stage design under proportional hazards assumption which allows exact calculations to be made for both design and analysis. Lin et al. (1996) and Wu and Xiong (2014) proposed multiple-stage designs. For evaluating survival distribution using one-sample log-rank test, Finkelstein et al. (2003), Sun et al. (2011), Wu (2014, 2015) and Schmidt et al. (2015) developed sample size formulae for single-stage design. Kwak and Jung (2013) proposed an optimal two-stage design under the exponential distribution. Belin et al. (2017) developed an optimal two-stage design with restricted follow-up under the exponential model. Wu et al. (2019), Shan and Zhang (2019) and Shan (2020) extended Belin's optimal two-stage design to other parametric survival distributions by using the exact variance estimate. Owzar and Jung (2008) considered the use of the median of a time-to-event variable as the primary endpoint to design phase II studies. Recently, Chu et al. (2020) proposed sample size calculation for single-arm phase II immunotherapy trials with long-term survivors and random delayed treatment effect.

Several Bayesian single-arm phase II designs have also been proposed for the time-to-event endpoints. Zhao et al. (2012) developed a two-stage phase II design with a survival endpoint based on the Bayes risk, assuming that time-to-event follows the Weibull distribution. Cotterill and Whitehead (215) developed a Bayesian single-arm trial design under the Weibull model too. Thall et al. (2005) proposed a Bayesian phase II design that makes go/no-go decision based on the posterior probability of the mean event time. Recently, Zhou et al. (2020) proposed a Bayesian optimal phase II design with time-to-event endpoint. Both Thall and Zhou's designs assumed time-to-event follows an exponential distribution.

Software for single-arm phase II survival trial design is very limited. The STPlan software version 4.5 and the web-based software offered by the SWOG provide sample size and power calculation for single-arm phase II trial designs with time-to-event endpoints, but both sample size calculations are limited to the exponential distribution only.

1.8   Scope and Motivation

Advances in the understanding of tumor biology and increasing knowledge of the molecular basis of anti-tumor immune response have led to an increase in molecularly targeted and immunotherapy clinical trials (Simon, 2016; Emens et al., 2016). For phase II trials of largely these cytostatic therapies where antitumor activity based on tumor response (tumor shrinkage) can't be directly hypothesized, instead of time-to-event endpoints (e.g., PFS) are more appropriate for such trial designs (Seymour et al., 2010). However, the resources including statistical methods and software are very limited. Furthermore, educational materials for both statistician and clinical trial investigators are needed. Reflecting this, the goal of this book is to provide a comprehensive summary of the most commonly used methods for single-arm phase II trial design with time-to-event endpoints. Some materials in this book are appearing for the first time. I hope this book will serve as a reference book for the researchers conducting clinical trials and for statisticians designing early phase clinical trials for pharmaceutical industries and cancer research institutes. It may also serve as part of a textbook for graduate students in the biostatistics field who wish to learn clinical trial design and sample size calculation for early phase clinical trials.

All sample size calculations for single-stage designs or two-stage procedures are implemented in R codes. Real examples from cancer clinical trials are used to illustrate the trial designs. Because most of the test statistics for single-arm phase II trials are not available in commercial software, all test statistics presented in the book are implemented in R codes and illustrated by data analysis.

2

Phase II Trial Design under Parametric Model

2.1 Introduction

The methods for single-arm phase II trial design with time-to-event endpoints are available from free softwares. The STPlan software version 4.5 (Brown et al., 2010) and the web-based software offered by the Southwest Oncology Group (SWOG) provide sample size and power calculation for single-arm phase II trial designs (https://stattools.crab.org/), but both sample size calculations are limited to the exponential distribution only. Besides, the STPlan didn't specify what test statistics were used for the sample size calculation which is a problem for the trial design because different test statistics will result in different study powers. The help document of SWOG online software indicates that their sample size calculation is derived from a cumulative hazard estimate based test statistic. However, it is well known that cumulative hazard estimate is restricted to nonnegative values and its asymptotic normality may not be satisfied, particularly when sample size is small for phase II trials. Furthermore, the exponential model has a constant hazard rate during the entire study period, and this may be invalid in a relatively longer follow-up duration of the trial with a time-to-event endpoint. To overcome these drawbacks, in this chapter, we will discuss single-arm phase II trial design under a more flexible parametric distribution, the Weibull distribution.

2.2 Weibull Model

We start with a parametric maximum likelihood estimate (MLE) test under the Weibull model because it is simple and its asymptotic distribution is easy to be derived. Assume that failure time variable T of a subject follows the Weibull distribution with a shape parameter κ and a scale parameter b. That is, T has a survival distribution function

$$S(t) = e^{-(\frac{t}{b})^{\kappa}}$$

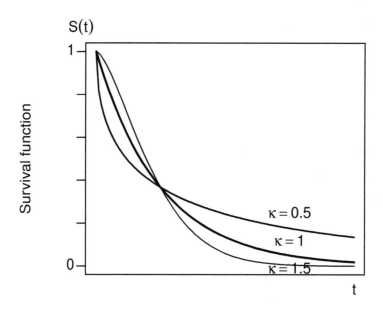

FIGURE 2.1: Weibull survival distribution functions with different shape parameters.

and a hazard function

$$\lambda(t) = \kappa b \left(\frac{t}{b}\right)^{\kappa - 1}.$$

The shape parameter κ indicates the cases of increasing ($\kappa > 1$), constant ($\kappa = 1$) or decreasing ($\kappa < 1$) hazard functions (Figures 2.1 and 2.2).

In a survival trial, the median survival time is an intuitive endpoint for clinicians. The median survival time for the Weibull distribution can be calculated as $m = b\{\log(2)\}^{1/\kappa}$ and the Weibull survival distribution can be expressed as

$$S(t) = e^{-\log(2)(\frac{t}{m})^{\kappa}}.$$

To design a single-arm phase II trial with a survival endpoint, increasing median survival time is always clinically meaningful for the investigators. For example, the median survival time of the standard of care is $m_0 = 6$ (months); an investigator wants to test the hypothesis to see if the new treatment can

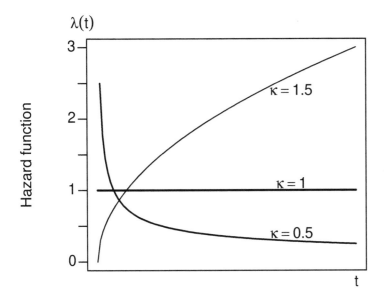

FIGURE 2.2: Weibull hazard functions with different shape parameters.

increase the median survival. Therefore the trial is designed for testing the following hypothesis

$$H_0 : m \le m_0 \quad vs \quad H_1 : m > m_0 \tag{2.1}$$

and powered at an alternative median survival time $m_1(> m_0)$ of the new treatment, e.g., $m_1 = 9$ (months). To derive the test statistics, we assume that the shape parameter κ is known or can be estimated from historical or literature data. For notational convenience, we convert the scale parameter to a hazard parameter (or rate parameter) $\lambda = \log(2)/m^\kappa$, and the corresponding survival distribution is $S(t) = e^{-\lambda t^\kappa}$.

2.3   Log Transformed MLE Test

Suppose during the accrual phase of the trial, n patients are enrolled in the study. Assuming that $\{T_i, i = 1, \ldots, n\}$ are independent failure times which follow the Weibull distribution $S(t) = e^{-\lambda t^\kappa}$, where the shape parameter κ is assumed a known constant. Assuming that $\{C_i, i = 1, \ldots, n\}$ are independent censoring times, and $\{T_i\}$ and $\{C_i\}$ are independent, then, the observed failure times and failure indicator are

$$X_i = \min(T_i, C_i) \quad \text{and} \quad \Delta_i = I(T_i \le C_i), \quad i = 1, \cdots, n,$$

where $I(\cdot)$ is an indicator function. On the basis of the observed data $\{X_i, \Delta_i, i = 1, \ldots, n\}$, the likelihood function of λ is given by

$$L(\lambda) = \lambda^D e^{-\lambda V},$$

and the log-likelihood function is

$$\ell(\lambda) = D \log(\lambda) - \lambda V,$$

where $D = \sum_{i=1}^{n} \Delta_i$ is the total observed number of failures and $V = \sum_{i=1}^{n} X_i^\kappa$ is the total amount of observed failure times rescaled by shape parameter κ. The maximum likelihood estimate (MLE) of parameter λ is obtained by solving the score equation $\ell'(\lambda) = d\ell(\lambda)/d\lambda = 0$, which is given by

$$\hat{\lambda} = \frac{D}{V}.$$

Figure 2.3 illustrates the log-likelihood function and maximum likelihood estimate (MLE).

As the distribution of $\hat{\lambda}$ is often skewed, we take a log transformation of $\hat{\lambda}$. Using delta-method, an approximate variance estimate of $\log(\hat{\lambda})$ is $1/D$. Thus,

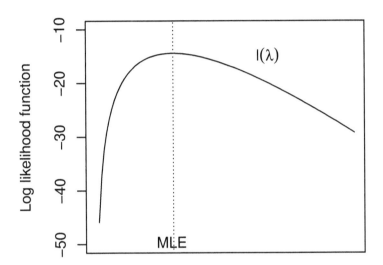

FIGURE 2.3: Log-likelihood function and maximum likelihood estimate.

under null hypothesis $H_0 : m = m_0$, the standardized Z_1 test is approximately standard normal distributed; that is,

$$Z_1 = \frac{\log(\lambda_0) - \log(\hat{\lambda})}{(1/D)^{1/2}} \sim N(0,1), \tag{2.2}$$

where $\lambda_0 = \log(2)/m_0^{\kappa}$. A larger value of Z_1 indicates evidence against the null hypothesis. Thus, we reject the null hypothesis H_0 if $Z_1 > z_{1-\alpha}$, where $z_{1-\alpha} = \Phi^{-1}(1-\alpha)$ and $\Phi(\cdot)$ is the cumulative distribution function (CDF) of standard normal.

The power $1 - \beta$ of the test statistic Z_1 under the alternative $H_1 : \lambda = \lambda_1$ satisfies the following equations:

$$\begin{aligned}
1 - \beta &= P(Z_1 > z_{1-\alpha}|H_1) \\
&= P\left(\frac{\log(\lambda_0) - \log(\hat{\lambda})}{(1/D)^{1/2}} > z_{1-\alpha}|H_1\right) \\
&\simeq P\left(\frac{\log(\lambda_1) - \log(\hat{\lambda})}{(1/D)^{1/2}} > z_{1-\alpha} - d^{1/2}\log(\lambda_0/\lambda_1)|H_1\right) \\
&\simeq \Phi\{-d^{1/2}\kappa\log(R) - z_{1-\alpha}\}, \tag{2.3}
\end{aligned}$$

where $d = E(D|H_1)$ is the total expected number of events under the alternative and

$$R = \frac{m_0}{m_1} = \left(\frac{\lambda_1}{\lambda_0}\right)^{1/\kappa}$$

is the median ratio. Solving d from equation (2.3), the total expected number of events is given by

$$d = \frac{(z_{1-\alpha} + z_{1-\beta})^2}{[\kappa\log(R)]^2} = \frac{(z_{1-\alpha} + z_{1-\beta})^2}{[\log(\delta)]^2}, \tag{2.4}$$

where $\delta = R^{\kappa} = \lambda_1/\lambda_0$ is the hazard ratio. It is easy to see $d = np$, where $p = P(\Delta = 1|H_1)$ is the probability of a patient having an event (failure) during the study (under the alternative). Thus, total sample size required for the study can be calculated as

$$n = \frac{d}{p} = \frac{(z_{1-\alpha} + z_{1-\beta})^2}{p[\kappa\log(R)]^2} = \frac{(z_{1-\alpha} + z_{1-\beta})^2}{p[\log(\delta)]^2}. \tag{2.5}$$

2.4 Quadratic Transformed MLE Test

Sprott (1973) showed that the asymptotic normality of a quadratic transformed MLE test is expected to be more accurate than the log transformed test. Thus, the following quadratic transformed MLE test Z_2 (Sprott, 1973; and Lawless, 1982) can be used to test the hypothesis:

$$Z_2 = \frac{\phi_0 - \hat{\phi}}{(\hat{\phi}^2/9D)^{1/2}}, \qquad (2.6)$$

where $\phi_0 = \lambda_0^{1/3}$, $\hat{\phi} = \hat{\lambda}^{1/3}$, and $\hat{\lambda} = D/V$, with $D = \sum_{i=1}^{n} \Delta_i$ and $V = \sum_{i=1}^{n} X_i^{\kappa}$. Under null hypothesis, the quadratic transformed MLE test Z_2 is approximately standard normal distributed. A larger value of Z_2 indicates evidence against the null hypothesis. Thus, we reject the null hypothesis H_0 if $Z_2 > z_{1-\alpha}$.

As $\hat{\phi} \to \phi_1 = \lambda_1^{1/3}$ under the alternative $H_1 : \lambda = \lambda_1$, the power $1 - \beta$ of the Z_2 test under the alternative $\lambda = \lambda_1$ satisfies the following:

$$
\begin{aligned}
1 - \beta &= P(Z_2 > z_{1-\alpha}|H_1) \\
&= P\left(\frac{\phi_0 - \hat{\phi}}{(\hat{\phi}^2/9D)^{1/2}} > z_{1-\alpha}|H_1\right) \\
&\simeq \Phi\left(\frac{\phi_1 - \phi_0}{(\phi_1^2/9d)^{1/2}} - z_{1-\alpha}\right), \qquad (2.7)
\end{aligned}
$$

where $d = E(D|H_1) = nP(\Delta = 1|H_1)$ is the expected total number of events of the new treatment. Solving d from equation (2.7), the total number of events required for the test statistic Z_2 can be calculated by

$$d = \frac{(z_{1-\alpha} + z_{1-\beta})^2}{9(\delta^{-1/3} - 1)^2}. \qquad (2.8)$$

and total sample size is given by

$$n = \frac{(z_{1-\alpha} + z_{1-\beta})^2}{9(R^{-\kappa/3} - 1)^2 p} = \frac{(z_{1-\alpha} + z_{1-\beta})^2}{9(\delta^{-1/3} - 1)^2 p}, \qquad (2.9)$$

where $R = m_0/m_1$ is the median ratio and $\delta = \lambda_1/\lambda_0 = R^{\kappa}$ is the hazard ratio, and $p = P(\Delta = 1|H_1)$ is the failure probability (e.g., disease progression) under the alternative hypothesis.

2.5 Sample Size Calculation

To calculate the sample size using formula (2.5) or (2.9), we have to calculate the failure probability p under the alternative hypothesis. Under the independent censorship assumption, that is, where failure time T and censoring time

C are independent, we have

$$
\begin{aligned}
p &= P(T < C|H_1) \\
&= \int_0^\infty P(T < C|T = t)f_1(t)dt \\
&= \int_0^\infty P(C > t)f_1(t)dt \\
&= \int_0^\infty S_1(t)\lambda_1(t)G(t)dt, \quad\quad\quad (2.10)
\end{aligned}
$$

where $f_1(t)$, $S_1(t)$ and $\lambda_1(t)$ are the density, survival and hazard function of T under the alternative and $G(t) = P(C > t)$ is the survival function of censoring time. We assume that subjects are uniformly recruited over an accrual period t_a and all subjects are followed until the end of the study with study duration $\tau = t_a + t_f$, where t_f is the follow-up time, a period from the last subject enrolled in the study to the end of the study. We further assume that no subjects are lost to follow-up during the trial. Let A be the accrual time of a subject, then, A follows a uniform distribution on the interval $[0, t_a]$ and the censoring is only administrative censoring; that is, a subject is censored if the subject does not have an event at the end of the trial. Therefore, the censoring time $C = \tau - A$ follows an uniform distribution on the interval $[t_f, t_a + t_f]$ and the survival distribution of the censoring time C is given by

$$
G(t) = \begin{cases} 1 & \text{if } 0 < t \le t_f \\ \frac{\tau - t}{t_a} & \text{if } t_f < t < \tau \\ 0 & \text{if } t \ge \tau \end{cases}
$$

Using formula (2.10), the failure probability of p can be calculated as follows:

$$
p = 1 - \frac{1}{t_a} \int_{t_f}^{t_a + t_f} e^{-\log(2)\left(\frac{t}{m_1}\right)^\kappa} dt,
$$

where the integration can be calculated using R function 'integrate' (derivation see Appendix A).

2.6   Accrual Duration Calculation

When designing an actual trial with a given accrual time t_a, calculating the sample size is often impractical because we may not be able to enroll the planned number of patients within the given accrual duration. It is more practical to design the study by starting with the given accrual rate r, calculating the required accrual time t_a. Using the sample size formula (2.9), we can define a root function of the accrual time t_a as follows:

$$\text{root}(t_a) = rt_a - \frac{(z_{1-\alpha} + z_{1-\beta})^2}{9(\delta^{-1/3} - 1)^2 p(t_a)}. \tag{2.11}$$

Now the accrual time t_a can be obtained by solving the root equation $\text{root}(t_a) = 0$ numerically in R using the 'uniroot' function. The total sample size required for the study is approximately $n = [rt_a]^+$, where $[x]^+$ denotes the smallest integer greater than x.

2.7 Comparison

To compare the performance of the two test statistics: the log-transformed test Z_1 and quadratic transformed test Z_2 in a small sample, we calculate the sample size under the following parameter configurations: median survival time under null hypothesis $m_0 = 5$ (months); median survival time under alternative $m_1 = 7$ to 10 by 1 (month); shape parameter of the Weibull distribution set to be $\kappa = 0.8, 1$ and 1.2; type I error rate $\alpha = 0.05$ and power of 80%. Sample sizes for two test statistics are calculated using formulae (2.5) and (2.9). Empirical type I error and power were estimated from 10,000 simulated trials. Results were recorded in Table 2.1. From simulation results, we have the following observations: a) The Z_1 test requires larger sample size or number of events than that of Z_2 test, b) The empirical type I error of Z_1 test is always below the nominal level 5% which indicates that the Z_1 test is a conservative test, c) The empirical power of Z_1 test is always larger than the nominal level of 80% which indicates that sample size based on formula (2.5) is overestimated. In contrast, the empirical type I error of Z_2 test is close to the nominal level 5% which indicates that the Z_2 preserves type I error well. Furthermore, the empirical power of Z_2 test is also close to the nominal level 80%, which shows that the sample size formula (2.9) of Z_2 test provides accurate sample size estimation. The Z_2 test requires smaller sample size than that of the Z_1 test, which indicates that the Z_2 is a more powerful test than that of the Z_1 test. Therefore we recommend using Z_2 test for the trial design under the Weibull model. Table 2.2 showed that the sample size depends on the underlying survival distribution, accrual distribution and length of follow-up. However, the number of events does not depend on the underlying accrual and survival distributions, which are often difficult to specify correctly in the design stage.

TABLE 2.1: Sample sizes (n) and number of events (d) were calculated under Weibull distribution with median survival time under null $m_0 = 5$ (months) and median survival time under alternative $m_1 = 7, 8, 9$ and 10 (months) for the two tests Z_1 and Z_2. The nominal type I error and power are set to be 0.05 and 80%, respectively. The corresponding empirical type I error $\hat{\alpha}$ (%) and empirical power (EP) (%) were simulated based on 10,000 simulation runs.

Test	m_1	$\kappa = 0.8$				$\kappa = 1$				$\kappa = 1.2$			
		d	n	$\hat{\alpha}$	EP	d	n	$\hat{\alpha}$	EP	d	n	$\hat{\alpha}$	EP
Z_1	7	86	145	4.3	81.4	55	90	4.3	82.0	38	60	4.0	81.8
	8	44	80	4.2	82.5	28	50	4.2	82.8	20	34	4.0	82.8
	9	28	54	4.2	83.3	18	35	3.9	84.5	13	24	3.8	84.3
	10	21	41	4.0	83.7	13	27	4.0	85.3	9	19	3.9	86.9
Z_2	7	78	133	5.0	80.0	49	80	4.8	79.4	34	52	4.9	79.3
	8	39	70	4.5	79.8	24	43	4.9	80.0	17	28	4.7	79.4
	9	24	46	5.2	79.9	15	28	4.6	79.5	10	19	4.8	80.0
	10	17	34	4.4	79.9	11	21	4.9	79.8	7	14	4.8	79.8

TABLE 2.2: Relationship between sample sizes (n), number of events (d) and follow-up and accrual duration for the two tests Z_1 and Z_2. Sample size and number of events are calculated under exponential model.

	Design			Z_1		Z_2	
m_0	m_1	t_a	t_f	d	n	d	n
5	7	10	5	86	145	78	133
5	7	10	10	86	120	78	110
5	7	10	15	86	108	78	99
5	7	10	20	86	101	78	92
5	7	20	5	86	124	78	113
5	7	20	10	86	110	78	100
5	7	20	15	86	102	78	93
5	7	20	20	86	97	78	89

TABLE 2.3: Time-to-disease relapse for patients with recurrent or refractory non-CNS rhabdoid tumors that were treated with conventional chemotherapy.

3.241	0.559	0.830	0.156	0.148	0.414	1.181	0.827	1.501	1.044	1.556	1.225	0.293

2.8 Study Design and Data Analysis

Example 1 *Trial for Rhabdoid Tumors*

Rhabdoid tumors are aggressive pediatric malignancies with a poor prognosis. Over the past 5-year period, St. Jude Children's Research Hospital had enrolled 13 pediatric patients with recurrent or refractory non-CNS rhabdoid tumors that were treated with conventional chemotherapy. All 13 patients had relapse within 3.25 years (see Table 2.3). The Weibull distribution $S(t) = e^{-(t/b)^\kappa}$ was fitted to the data by using R code 'Fit', which resulted in an estimate of the shape parameter $\kappa = 1.32$ and scale parameter $b = 1.087$. The median relapse-free survival (RFS) time is $m = b\{\log(2)\}^{1/\kappa} = 0.823$ years (see Figure 2.5). The Kaplan-Meier estimate of 1-year RFS was 46% with standard error (se) of 14%.

Suppose that we wish to design a new trial and consider that the molecular agent alisertib is not worthy of further evaluation if the median RFS is at most 0.8 year and it is promising if the median RFS is at least 1.2 year. For the trial design, we assume that the RFS time follows the Weibull distribution $S(t) = e^{-\lambda t^\kappa}$, where $\lambda = \log(2)/m^\kappa$ and $\kappa = 1.32$; uniform accrual with accrual period $t_a = 2$ years; follow-up period $t_f = 1$ year; no loss to follow-up. With 80% power and 5% one-sided type I error rate, the required sample size calculated from formulae (2.5) and (2.9) are 31 and 26, respectively. If annual accrual rate is 13 patients, using accrual duration formula (2.11), the calculated the study duration is 1.95 years and required sample size is $1.95 \times 13 \simeq 26$ patients which is the same as the sample size calculated using formula (2.9). We estimated empirical power for the study design using Z_2 test ($n = 26$) based on 100,000 simulated trials which give the empirical power of 78.7% close to the nominal level 80%.

For illustration of data analysis after the trial was conducted, we use a random sample data generated from simulation to calculate the p-values of Z_1 and Z_2 for testing hypothesis $H_0 : m_0 = 0.8$ using R code 'Test' which gives the p-values 0.0093 and 0.0049 for Z_1 and Z_2, respectively. Thus, both tests reject the null hypothesis $H_0 : m_0 = 0.8$ and support the alternative hypothesis $H_1 : m_1 = 1.2$ and we can conclude that the new treatment is promising.

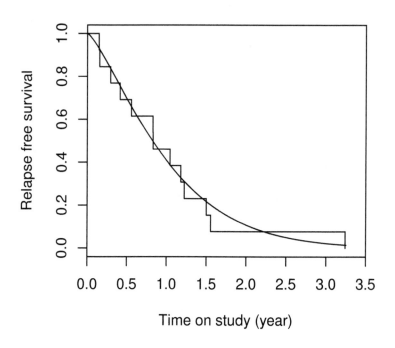

FIGURE 2.4: Kaplan-Meier curve (step function) and fitted Weibull distribution (smooth curve).

2.9   R Code

In this section, R codes are provided for fitting the Weibull survival distribution by code 'Fit', for sample size calculation by code 'Size', for study duration calculation by code 'Duration', for power simulation by code 'pow' and for data analysis after the trial by code 'Test'.

```
################## Fitting Weibull Distribution ############################
### data is a data frame with time variable which is observed failure time  ###
### and status is failure indicator, 1-failure, 0-censoring, x is landmark  ###
### time point at which the survival probability is estimated.              ###
###########################################################################
time=c(3.241, 0.559, 0.830, 0.156, 0.148, 0.414, 1.181,
       0.827, 1.501, 1.044, 1.556, 1.225, 0.293)
status=rep(1,13)
data=data.frame(time=time, status=status)
Fit=function(x,data)
{ fitWB=survreg(Surv(time,status)~1,data = data,dist="weibull")
  scale <- as.numeric(exp(fitWB$coef))
  shape <- 1/(fitWB$scale)
  median <- scale*(log(2))^(1/shape)
  fitKM=survfit(Surv(time, status)~1, data = data)
  std=tail(summary(fitKM)$std[summary(fitKM)$time<=x],1)
  S.hat=tail(summary(fitKM)$surv[summary(fitKM)$time<=x],1)
  ans=list(shape=round(shape,4),scale=round(scale,4),
    median=round(median,4),S.hat=round(S.hat,4),std=round(std,4))
  return(ans)}
Fit(x=1, data=data)
$shape  ## Weibull shape parameter
[1] 1.3152
$scale  ## Weibull scale parameter
[1] 1.0869
$median ## median survival time
[1] 0.8226
$S.hat  ## KM estimate of survival prob at x
[1] 0.4615
$std    ## stanard error of the KM estimate
[1] 0.1383

######################### Sample Size Calculation ##########################
### kappa is the Weibull shape parameter; m0 and m1 are median survival of  ###
### null and alternative; alpha and beta are the type I and type II errors  ###
### and power=1-beta; ta and tf are the accrual and follow-up time;         ###
###########################################################################
Size=function(kappa, m0, m1, ta, tf, alpha, beta)
{ tau=ta+tf
  lambda0=log(2)/m0^kappa
  lambda1=log(2)/m1^kappa
  delta=lambda1/lambda0
  z0=qnorm(1-alpha); z1=qnorm(1-beta)
  S1=function(t){exp(-lambda1*t^kappa)}
  p=1-integrate(S1, tf, tau)$value/ta
  nZ1=(z0+z1)^2/(log(delta)^2*p)
  nZ2=(z0+z1)^2/((delta^(-1/3)-1)^2*9*p)
```

```
      ans=list(c(nZ1=ceiling(nZ1),nZ2=ceiling(nZ2)))
      return(ans)}
Size(kappa=1.32, m0=0.8, m1=1.2, ta=2, tf=1, alpha=0.05, beta=0.2)
nZ1 nZ2
31  26


######################### Duration Calculation #############################
### kappa is the Weibull shape parameter; m0 and m1 are median survival of  ###
### null and alternative; alpha and beta are the type I and type II errors  ###
### and power=1-beta; r is the accrual rate; tf is follow-up time;          ###
#############################################################################
Duration=function(kappa, m0, m1, alpha, beta, r, tf)
{
 root=function(ta)
 {lambda0=log(2)/m0^kappa; lambda1=log(2)/m1^kappa; tau=ta+tf
  S1=function(t){exp(-lambda1*t^kappa)}
  p=1-integrate(S1, tf, tau)$value/ta
  z0=qnorm(1-alpha); z1=qnorm(1-beta); delta=lambda1/lambda0
  ans=r*ta-(z0+z1)^2/(9*(delta^(-1/3)-1)^2*p)}
  ta=uniroot(root, lower=0.1, upper=200)$root
  n=ceiling(r*ta); ta=round(ta, 2);
  ans=list(c(ta=ta, n=n))
  return(ans)}
Duration(kappa=1.32,m0=0.8,m1=1.2,alpha=0.05,beta=0.2,r=13,tf=1)
ta     n
1.95   26


################## Power and Type I Error Simulation #######################
### kappa is the Weibull shape parameter; m0 and m1 are median survival of  ###
### null and alternative; alpha and beta are the type I and type II errors  ###
### and power=1-beta; r is the accrual rate; tf is follow-up time;          ###
#############################################################################
## 1. Empirical power of test Z1 ###
pow=function(kappa, m0, m1, ta, tf, alpha, n)
{lambda0=log(2)/m0^kappa
 lambda1=log(2)/m1^kappa
 rho=lambda1^(1/kappa)    # 1/rho is Weibull scale parameter ##
 shape=kappa
 scale=1/rho
 z0=qnorm(1-alpha)
 tau=ta+tf
 s=0
 N=100000
 set.seed(852)
 for (i in 1:N)
 {t=rweibull(n, shape, scale) # generate random sample under alternative #
  u=runif(n, 0, ta)
  delta = as.numeric(t<ta+tf-u) # only consider administrative censoring #
  d=sum(delta)
  x=pmin(t, ta+tf-u)
  V=sum(x^kappa)
  lambda.hat=d/V
  Z=(log(lambda0)-log(lambda.hat))/(1/d)^(1/2)
  if (Z>z0) (s=s+1)
 }
 pow=round(s/N,3)
 return(pow)
```

```
}
pow(kappa=1.32, m0=0.8, m1=1.2, ta=2, tf=1, alpha=0.05, n=31)
[1] 0.822
pow(kappa=1.32, m0=0.8, m1=0.8, ta=2, tf=1, alpha=0.05, n=31)
[1] 0.039

## 2. Empirical power of test Z2 ###
pow=function(kappa, m0, m1, ta, tf, alpha, n)
{lambda0=log(2)/m0^kappa
 lambda1=log(2)/m1^kappa
 rho=lambda1^(1/kappa)    # 1/rho is Weibull scale parameter ##
 shape=kappa
 scale=1/rho
 z0=qnorm(1-alpha)
 tau=ta+tf
 s=0
 N=100000
 set.seed(852)
 for (i in 1:N)
 {t=rweibull(n, shape, scale)# generate random sample under alternative #
  u=runif(n, 0, ta)
  delta = as.numeric(t<ta+tf-u)# only consider administrative censoring #
  d=sum(delta)
  x=pmin(t, ta+tf-u)
  V=sum(x^kappa)
  lambda.hat=d/V
  phi0=lambda0^(1/3)
  phi.hat=lambda.hat^(1/3)
  Z=(phi0-phi.hat)/(phi.hat^2/(9*d))^(1/2)
  if (Z>z0) (s=s+1)
 }
 pow=round(s/N,3)
 return(pow)
}
pow(kappa=1.32, m0=0.8, m1=1.2, ta=2, tf=1, alpha=0.05, n=26)
[1] 0.787
pow(kappa=1.32, m0=0.8, m1=0.8, ta=2, tf=1, alpha=0.05, n=26)
[1] 0.047


############### Example of data analysis using simulated data  ###############
### kappa is the Weibull shape parameter; m0 and m1 are median survival of  ###
### null and alternative; ta is the accrual duration; tf is follow-up time; ###
### n is sample size. 'Test' calculates the p-value for testing hypothesis  ###
### H0: m=m0; data is generated under the alternative H1:m=m1 hypothesis.    ###
##############################################################################
Test=function(kappa, m0, m1, ta, tf, n)
{lambda0=log(2)/m0^kappa
 lambda1=log(2)/m1^kappa
 shape=kappa
 scale=1/lambda1^(1/kappa)
 tau=ta+tf
 set.seed(825)
 t=rweibull(n, shape, scale)
 u=runif(n, 0, ta)
 delta = as.numeric(t<ta+tf-u)
 d=sum(delta)
 x=pmin(t, ta+tf-u)
```

```
V=sum(x^kappa)
lambda.hat=d/V
phi0=lambda0^(1/3)
phi.hat=lambda.hat^(1/3)
Z1=(log(lambda0)-log(lambda.hat))/(1/d)^(1/2)
p.Z1=1-pnorm(Z1)
Z2=(phi0-phi.hat)/(phi.hat^2/(9*d))^(1/2)
p.Z2=1-pnorm(Z2)
ans=list(p.Z1=round(p.Z1,4),p.Z2=round(p.Z2, 4))
return(ans)
}
Test(kappa=1.32, m0=0.8, m1=1.2, ta=2, tf=1, n=26)
$p.Z1
[1] 0.0093
$p.Z2
[1] 0.0049
```

3

One-Stage Design Evaluating Survival Probabilities

3.1 Introduction

For a single-arm phase II oncology trial, the primary endpoint is often the survival probability at a clinically meaningful landmark time point x. Let $S_0(x)$ be the survival probability at landmark time point x which investigators are no longer interested in and $S(x)$ be the survival probability at landmark time point x of the new treatment, then, the trial design can be based on testing the following one-sided hypothesis:

$$H_0 : S(x) \leq S_0(x) \quad vs \quad H_1 : S(x) > S_0(x),$$

at landmark time point x; or equivalent to test the following hypothesis for the cumulative hazard function $\Lambda(x) = -\log S(x)$,

$$H_0 : \Lambda(x) \geq \Lambda_0(x) \quad vs \quad H_1 : \Lambda(x) < \Lambda_0(x), \tag{3.1}$$

at landmark time point x, where $\Lambda_0(x) = -\log S_0(x)$. Figure 3.1 illustrates the hypothesis testing of survival probabilities at a fixed landmark time point x.

3.2 Nelson-Aalen Estimate Based Tests

The single-arm phase II trial with endpoint of survival probability at a fixed landmark time point x can be designed using Nelson-Aalen estimate based test statistics proposed by Lin et al. (1996). The test statistics are introduced in the next section and their asymptotic distributions are derived for the trial design.

3.2.1 Test Statistics

Suppose during the accrual phase of the trial, n subjects are enrolled in the study. Let T_i and C_i denote, respectively, the failure time and censor-

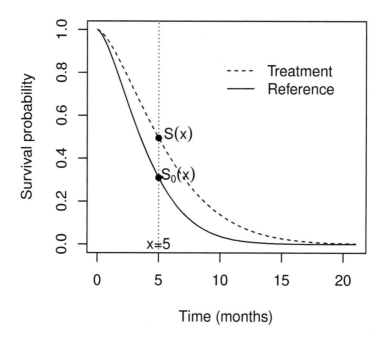

FIGURE 3.1: Hypothesis testing of survival probabilities at a fixed landmark time point x.

ing time of the ith subject, with both being measured from the time of study entry of this subject. We assume that the failure time T_i is independent of the censoring time C_i, and $\{(T_i, C_i); i = 1, \cdots, n\}$ are independent and identically distributed. When the data are examined at the end of the study, we observe the time to failure $X_i = T_i \wedge C_i$ and failure indicator $\Delta_i = I(T_i \leq C_i), i = 1, \cdots, n$, where $x \wedge y = \min(x, y)$. Further define failure process $N_i(t) = \Delta_i I(X_i \leq t)$ and at-risk process $Y_i(t) = I(X_i \geq t)$, and $\bar{N}(t) = \sum_{i=1}^{n} N_i(t)$ and $\bar{Y}(t) = \sum_{i=1}^{n} Y_i(t)$. The hazard function $\lambda(t)$ of failure time can be estimated by

$$\hat{\lambda}(t) = \frac{d\bar{N}(t)}{\bar{Y}(t)},$$

where $d\bar{N}(t)$ is defined as $\bar{N}((t + dt)^-) - \bar{N}(t^-)$, the number of failures that occur in the interval $[t, t + dt)$, where $\bar{N}(t^-) = \lim_{\Delta t \to 0} \bar{N}(t - \Delta t)$ (left limit) is the value of the process at the instant before t. Therefore, the Nelson-Aalen estimate of the cumulative hazard function $\Lambda(x) = \int_0^x \lambda(u) du$ is given by

$$\hat{\Lambda}(x) = \int_0^x \frac{d\bar{N}(u)}{\bar{Y}(u)}. \tag{3.2}$$

Under the null hypothesis, it is well known that the process (Fleming and Harrington, 1991; also see Appendix B))

$$\sqrt{n}\{\Lambda_0(x) - \hat{\Lambda}(x)\}$$

converges weakly to a normal distribution with mean zero and variance of

$$\sigma^2(x) = \int_0^x \frac{d\Lambda(u)}{\pi(u)} \tag{3.3}$$

where $\pi(u) = P(X_1 \geq u)$, and variance $\sigma^2(x)$ can be estimated by

$$\hat{\sigma}^2(x) = \int_0^x \frac{d\bar{N}(u)}{\bar{Y}^2(u)/n}. \tag{3.4}$$

However, the distribution of $\hat{\Lambda}(x)$ is often skewed, and it requires large sample sizes to maintain its type I error rate. This is partly because $\hat{\Lambda}(x)$ is restricted to nonnegative values. However, a logarithmic transformation of the Nelson-Aalen estimate $\log\{\hat{\Lambda}(x)\}$ takes the value over the entire real line, the asymptotic normality is expected to be more accurate. By delta-method, the variance estimate of $\log\{\hat{\Lambda}(x)\}$ is approximately given by

$$\widehat{\mathrm{Var}}\{\log \hat{\Lambda}(x)\} \simeq \frac{\hat{\sigma}^2(x)}{\hat{\Lambda}^2(x)}.$$

Therefore, we test the hypothesis (3.1) using the following standardized test statistic

$$Z_1(x) = \frac{\sqrt{n}\{\log \Lambda_0(x) - \log \hat{\Lambda}(x)\}\hat{\Lambda}(x)}{\hat{\sigma}(x)}. \tag{3.5}$$

which is asymptotically standard normal distributed under the null hypothesis. It is referred to as the log-log transformed test because of $\log \Lambda(x) = \log\{-\log S(x)\}$. As the better of the treatment effect results in a larger reduction of the cumulative hazard, we reject null hypothesis for a large observed value of $Z_1(x)$ test. Specifically, given type I error rate α, if $Z_1(x) > z_{1-\alpha}$, we reject the null hypothesis.

The second test statistic is constructed based on an arcsin-square root transformation on $\hat{S}(x)$. Using delta-method, it can be shown that

$$\widehat{\text{Var}}\left\{\arcsin\sqrt{\hat{S}(x)}\right\} \simeq \frac{\hat{S}(x)\hat{\sigma}^2(x)}{4\{1 - \hat{S}(x)\}}.$$

Therefore, the arcsin-square root transformed test is given by

$$Z_2(x) = \frac{\sqrt{n}\{\arcsin\sqrt{\hat{S}(x)} - \arcsin\sqrt{S_0(x)}\}}{\hat{\nu}(x)}, \tag{3.6}$$

where

$$\hat{\nu}^2(x) = \frac{\hat{S}(x)\hat{\sigma}^2(x)}{4\{1 - \hat{S}(x)\}}.$$

The third test statistic is defined based on the logit transformation, which is given by

$$Z_3(x) = \frac{\sqrt{n}\{\text{logit}\hat{S}(x) - \text{logit}S_0(x)\}\{1 - \hat{S}(x)\}}{\hat{\sigma}(x)}, \tag{3.7}$$

where $\hat{\sigma}^2(x)/\{1 - \hat{S}(x)\}^2$ is the asymptotic variance estimate of the logit$\hat{S}(x)$ obtained using delta-method.

By counting process integration of equations (3.2) and (3.4), the test statistics Z_1, Z_2 and Z_3 based on Nelson-Aalen estimate can be calculated by using the following two equations

$$\hat{\Lambda}(x) = \sum_{i=1}^{n} \frac{\Delta_i I(X_i \le x)}{\sum_{j=1}^{n} I(X_j \ge X_i)},$$

and

$$\hat{\sigma}^2(x) = n\sum_{i=1}^{n} \frac{\Delta_i I(X_i \le x)}{[\sum_{j=1}^{n} I(X_j \ge X_i)]^2}.$$

The Nelson-Aalen estimate for the survival function is given as $\hat{S}(x) = e^{-\hat{\Lambda}(x)}$, which is asymptotically equivalent to the Kaplan-Meier estimator.

3.2.2   Sample Size Formula

A key step in designing a clinical trial is to calculate the required minimal sample size constrained by type I error rate and power. If the study is interested in testing hypotheses (3.1), then given type I error rate α and power

$1 - \beta$ at the alternative hypothesis, sample size formula for fixed sample tests can be derived as follows.

The power of a test is the probability of rejecting the null hypothesis when the alternative is true. Hence, for fixed sample test, the power of the test statistic $Z_1(x)$ under the alternative hypothesis $H_1 : \Lambda(x) = \Lambda_1(x)$ is determined by

$$
\begin{aligned}
1 - \beta &= P(Z_1(x) > z_{1-\alpha}|H_1) \\
&= P\{\sqrt{n}\hat{\sigma}^{-1}(x)(\log \Lambda_0(x) - \log \hat{\Lambda}(x))\hat{\Lambda}(x) > z_{1-\alpha} \,|H_1\} \\
&\simeq \Phi\{\sqrt{n}\sigma_1^{-1}(x)(\log \Lambda_0(x) - \log \Lambda_1(x))\Lambda_1(x)) - z_{1-\alpha}\}, \quad (3.8)
\end{aligned}
$$

where $\sigma_1^2(x)$ is the asymptotic variance of $\hat{\Lambda}(x)$ at the alternative. Thus, sample size formula for the test statistic $Z_1(x)$ is given by

$$
n = \frac{(z_{1-\alpha} + z_{1-\beta})^2 \sigma_1^2(x)}{\{\log \Lambda_0(x) - \log \Lambda_1(x)\}^2 \Lambda_1^2(x)}. \qquad (3.9)
$$

Similar, sample size formula for $Z_2(x)$ is given by

$$
n = \frac{(z_{1-\alpha} + z_{1-\beta})^2 \nu_1^2(x)}{\{\arcsin \sqrt{S_1(x)} - \arcsin \sqrt{S_0(x)}\}^2}, \qquad (3.10)
$$

where

$$
\nu_1^2(x) = \frac{S_1(x)\sigma_1^2(x)}{4\{1 - S_1(x)\}},
$$

and sample size formula for $Z_3(x)$ is given by

$$
n = \frac{(z_{1-\alpha} + z_{1-\beta})^2 \sigma_1^2(x)}{\{\mathrm{logit}\, S_1(x) - \mathrm{logit}\, S_0(x)\}^2 \{1 - S_1(x)\}^2}. \qquad (3.11)
$$

3.2.3 Sample Size Calculation

To calculate the sample size n given in the above formulae, we have to calculate the asymptotic variance $\sigma_1^2(x)$ under the alternative hypothesis, which is given by (3.3). We assume that the survival distribution of censoring time is $G(u)$. Then, under the alternative, $\pi(u) = P(T_1 \wedge C_1 > u) = P(T_1 > u)P(C_1 > u) = S_1(u)G(u)$. Thus, by equation (3.3), we have

$$
\sigma_1^2(x) = \int_0^x \frac{\lambda_1(u)}{S_1(u)G(u)} du, \qquad (3.12)
$$

where $\lambda_1(\cdot)$ and $S_1(\cdot)$ are the hazard and survival distribution function at the alternative, respectively. Assuming subjects were recruited with a uniform distribution over the accrual period t_a and followed for an additional period of length t_f, then the study duration $\tau = t_a + t_f$. If there is no loss to follow-up, then censoring is only the administrative censoring. Therefore the survival distribution of censoring time is uniformly distributed on interval $[t_f, t_a + t_f]$ and survival function of $G(u)$ is given by

$$G(u) = \begin{cases} 1 & \text{if } 0 < u \le t_f \\ \frac{\tau - u}{t_a} & \text{if } t_f < u < \tau \\ 0 & \text{if } u \ge \tau \end{cases}$$

- When the length of follow-up is shorter than landmark time point, that is $t_f < x$, the variance can be calculated by

$$\sigma_1^2(x) = \int_0^x \frac{\lambda_1(u)}{S_1(u)G(u)}. \tag{3.13}$$

In this case, to calculate variance $\sigma_1^2(x)$, we need to know the hazard function $\lambda_1(u)$ and survival distribution function $S_1(\cdot)$ under the alternative hypothesis. Thus, the study design is not distribution-free even through the test statistic $Z_1(x)$ is a non-parametric test.

- When length of follow-up is greater than or equal to the landmark time point, that is $t_f \ge x$, then $G(u) = 1$ for any $0 < u < x \le t_f$; hence the equation (3.13) can be simplified as

$$\sigma_1^2(x) = \int_0^x \frac{\lambda_1(u)}{S_1(u)} du = \frac{1}{S_1(x)} - 1. \tag{3.14}$$

When there is no loss to follow-up and follow-up time $t_f \ge x$, sample size for a fixed sample test depends only on the single value $S_1(x)$, which is the survival probability at landmark time point x at the alternative hypothesis. Therefore, the study design is free of the underlying survival distribution. However when $t_f \le x$, the study design depends on the underlying survival distribution. We implemented the following two trial design scenarios for sample size calculation.

In the first scenario, we assume that the null and alternative survival distributions follow the same type of distributions, e.g., the null distribution is a Weibull distribution, then the alternative distribution is a Weibull distribution with the same shape parameter but different scale parameter. Sample size calculations are implemented for four commonly used survival distributions: Weibull (WB), gamma (GM), log-normal (LN) and log-logistic (LG) (see Table 3.1), where the shape parameters are assumed known (an estimate from historical or literature data) and the scale parameters are calculated from $S_0(x)$ and $S_1(x)$ for the null and alternative survival distributions, respectively. For example, for the Weibull distribution $S(x) = e^{-(x/b)^a}$, the scale parameters are solved to be $b_0 = x/\{-\log S_0(x)\}^{1/a}$ and $b_1 = x/\{-\log S_1(x)\}^{1/a}$ for the null and alternative survival distributions, respectively. Thus, the survival distributions are $S_0(x) = e^{-(x/b_0)^a}$ and $S_1(x) = e^{-(x/b_1)^a}$, for the null and alternative survival distributions, respectively. Therefore, in general the distribution under the alternative does not

TABLE 3.1: Four parametric survival distributions with shape parameter a and scale parameter b.

	Surv. function	Density	Cumu. hazard	Hazard
Dist.	$S(t)$	$f(t)$	$\Lambda(t)$	$h(t)$
WB	$e^{-(\frac{t}{b})^a}$	$(\frac{a}{b})(\frac{t}{b})^{a-1}e^{-(\frac{t}{b})^{a-1}}$	$(\frac{t}{b})^a$	$(\frac{a}{b})(\frac{t}{b})^{a-1}$
GM	$1 - I_a(\frac{t}{b})$	$\frac{(\frac{t}{b})^{a-1}e^{-\frac{t}{b}}}{\Gamma(a)b}$	$-\log S(t)$	$\frac{f(t)}{S(t)}$
LN	$1 - \Phi(\frac{\log t - b}{a})$	$\frac{1}{\sqrt{2\pi}at}e^{-\frac{(\log t - b)}{2a^2}}$	$-\log S(t)$	$\frac{f(t)}{S(t)}$
LG	$\frac{1}{1+(\frac{t}{b})^a}$	$\frac{(\frac{a}{b})(\frac{t}{b})^{a-1}}{[1+(\frac{t}{b})^a]^2}$	$\log(1 + (\frac{t}{b})^a)$	$\frac{(\frac{a}{b})(\frac{t}{b})^{a-1}}{1+(\frac{t}{b})^a}$

satisfy the proportional hazard assumption to the null distribution (except for the exponential distribution). The R code for sample size calculation in the chapter is implemented under this scenario.

In the second scenario, we adapt the proportional hazards assumption for the study design. We assume the null survival distribution is $S_0(x)$, which is one of the four parametric survival distributions. Given a hazard ratio δ (< 1), the alternative survival distribution, density function and cumulative hazard function are determined as $S_1(x) = [S_0(x)]^\delta$, $f_1(x) = \delta[S_0(x)]^{\delta-1}f_0(x)$ and $\Lambda_1(x) = -\delta \log S_0(x) = \delta\Lambda_0(x)$, respectively, where $f_0(x)$ and $\Lambda_0(x)$ are the density function and cumulative hazard function under the null hypothesis. The R code for the sample size calculation under this scenario can be obtained by simple modifications of the R code for the first scenario.

Remark 1: All three tests $Z_1(x)$, $Z_2(x)$ and $Z_3(x)$ are non-parametric tests which are distribution-free tests. However, the calculation of the asymptotic variance $\sigma_1^2(x)$ in their sample size formulae requires that we know the survival distribution under the alternative hypothesis when the landmark time point x is greater than the follow-up time t_f. However if $t_f \geq x$, the asymptotic variance $\sigma_1^2(x)$ depends on the value of $S_1(x)$ only and is independent of underlying survival distribution. Therefore the study design is distribution-free. Thus, when the underlying survival distribution under the alternative is difficult to specify, we recommend choosing the follow-up duration $t_f \geq x$ for the study design.

Remark 2: When a patient's follow-up time is longer than x unit time because the test statistics use the data only up to x unit time, thus, the study design and data analysis are the same as using the full follow-up data but truncated to x unit follow-up time.

Example 2 *Trial for Pancreatic Cancer*
Case and Morgan (2003) illustrated a single-arm phase II trial design to

access chemo-radiation treatment for patients with resectable pancreatic cancer. The 12-month survival probability of 35% or less on treatment is considered inefficient, while 12-month survival probability of 50% or greater would be worthwhile. To illustrate real trial data from this single-arm phase II trial, a sample of 68 survival times of experimental subjects were simulated by Whitehead (2014), where survival times follow an exponential distribution $S(t) = e^{-\lambda t}$ with rate parameter $\lambda = -\log S(12)/12 = 0.0578/12$. Assuming patients were uniformly recruited at a rate of 2 patients per month for 34 months, with final analysis at 46 months. There was no loss to follow-up, thus, censoring was administrative censoring only and a total of 20 observations were administratively censored. The full follow-up data were given in Table 3.2. To calculate the test statistics, we have to calculate the Nelson-Aalen estimate $\hat{\Lambda}(x)$ and its variance estimate $\hat{\sigma}^2$. Using R function 'coxph', we obtain $\hat{S}(12) = 0.547$ and standard error of $\hat{S}(12)$ is std.err=0.0601, or $\hat{\sigma}(12) = \sqrt{n} \times std.err/\hat{S}(12)$. The Nelson-Aalen estimate of survival curve is given in Figure 3.2. Using $\Lambda_0(12) = -\log(0.35)$ and $\hat{\Lambda}(12) = -\log\{\hat{S}(12)\} = -\log(0.547)$, the observed value of log-log transformed test is

$$Z_1(12) = \frac{\{\log(\Lambda_0(12)) - \log(\hat{\Lambda}(12))\}\hat{\Lambda}(12)}{\{std.err/\hat{S}(12)\}} = 3.0457$$

Similar, observed values of the arcsin square-root and logit transformed tests are $Z_2(12) = 3.3086$ and $Z_3(12) = 3.3346$.

```
########## Calculate Nelson-Aalen Estimate and Test Statistics #############
### S0 is the survival probability at landmark time point x under the null; ###
### data is the data frame with variable Time:time to death or the analysis ###
### and variable Cens: censoring indicator; Entry: time entry to the trial. ###
################################################################################
data=data.frame(Time=Time,Cens=Cens,Entry=Entry)
Test=function(S0,x,data){
  Time=data$Time; Cens=data$Cens
  fitNA=survfit(coxph(Surv(Time,Cens)~1), type = "aalen")
  std.err.NA=tail(summary(fitNA)$std[summary(fitNA)$time<=x],1)
  S.hat.NA=tail(summary(fitNA)$surv[summary(fitNA)$time<=x],1)
  Lam0=-log(S0); Lam.hat.NA=-log(S.hat.NA)
  Z1.NA=(log(Lam0)-log(Lam.hat.NA))*Lam.hat.NA/(std.err.NA/S.hat.NA)
  nu.hat.NA=std.err.NA^2/(4*S.hat.NA*(1-S.hat.NA))
  Z2.NA=(asin(sqrt(S.hat.NA))-asin(sqrt(S0)))/sqrt(nu.hat.NA)
  logit=function(t){log(t/(1-t))}
  Z3.NA=(logit(S.hat.NA)-logit(S0))*(1-S.hat.NA)/(std.err.NA/S.hat.NA)
  Z1.NA=round(Z1.NA,4)
  Z2.NA=round(Z2.NA,4)
  Z3.NA=round(Z3.NA,4)
  ans=c(Z1.NA=Z1.NA,Z2.NA=Z2.NA, Z3.NA=Z3.NA,
     S.hat.NA=round(S.hat.NA,4),std.err.NA=round(std.err.NA,4))
  return(ans)
}
Test(S0=0.35,x=12,data)
  Z1.NA      Z2.NA      Z3.NA    S.hat.NA std.err.NA
 3.0458     3.3086     3.3346     0.5475     0.0601
```

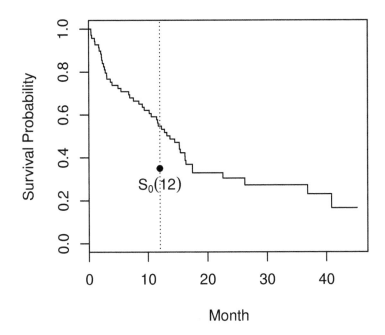

FIGURE 3.2: Nelson-Aalen survival curve for the full follow-up data and survival probability under the null hypothesis.

    To illustrate the data with restricted $x = 12$ months follow-up, the observed survival times and censoring indicators will be $X_i(t) = T_i \wedge x \wedge (t - A_i)^+$ and failure indicator $\Delta_i(t) = I(T_i \leq x \wedge (t - A_i)^+), i = 1, \cdots, n$, where A_i is the entry time of the ith subject which is uniformly distributed on interval $[0, 34]$ (months) and $t = 46$ is the final analysis time. We create this restricted follow-up data (given in Table 3.3) using the full follow-up data for illustration of trial design and data analysis. We obtained the same Nelson-Aalen estimate $\hat{\Lambda}(x)$ and its variance estimate $\hat{\sigma}^2$ by using this restricted follow-up data and hence the same observed values of the test statistics $Z_1(12), Z_2(12)$ and $Z_3(12)$.

    Sample size calculations are implemented for the case of $t_f > x$, thus the study design is free of the underlying survival distribution, accrual distribution or censoring distribution. We need only input the null and alternative hypothesis values of survival probabilities $S_0(x)$ and $S_1(x)$ at landmark time point x; and type I error rate α and type II error rate β (or power 1-β). Following R code 'Size' is implemented for the trial design in case follow-up time is

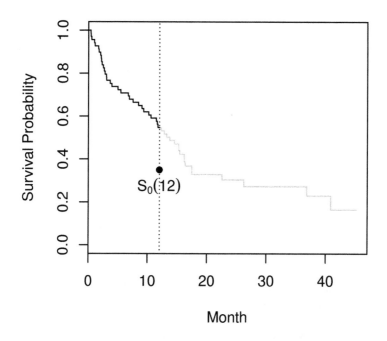

FIGURE 3.3: Nelson-Aalen survival curve for the data with restricted follow-up (black) and survival probability under the null hypothesis.

TABLE 3.2: Simulated full follow-up data ($n = 68$) used in the illustrative analysis: entry is time of entry to the trial (months), time is time to death or to the analysis (months); cens is censoring indicator (0=censored, 1=died).

Entry	Time	Cens	Entry	Time	Cens	Entry	Time	Cens
0.824	45.176	0	16.212	3.034	1	24.620	13.198	1
2.264	43.736	0	16.844	2.157	1	25.692	0.386	1
2.700	2.039	1	17.035	11.736	1	26.256	10.223	1
3.006	17.459	1	17.262	28.738	0	26.643	0.987	1
3.209	40.862	1	17.792	14.514	1	26.755	19.245	0
3.330	36.811	1	17.965	6.778	1	26.930	19.070	0
4.908	12.297	1	18.741	9.080	1	27.098	15.445	1
5.169	26.250	1	18.766	2.662	1	28.241	3.971	1
5.553	7.575	1	19.428	13.726	1	28.252	3.674	1
5.729	15.251	1	20.277	1.084	1	28.637	17.363	0
6.882	4.985	1	20.430	25.570	0	28.781	17.219	0
7.733	11.474	1	20.526	22.573	1	29.937	3.076	1
7.842	38.158	0	20.570	2.801	1	29.985	16.015	0
8.086	15.278	1	21.506	2.196	1	30.291	0.522	1
9.160	36.840	0	21.570	5.477	1	30.351	15.649	0
10.472	6.951	1	22.023	23.977	0	30.732	15.268	0
10.934	0.421	1	22.632	16.242	1	30.854	10.580	1
12.941	1.783	1	22.686	12.789	1	31.566	1.713	1
12.993	8.493	1	23.093	22.907	0	31.623	9.374	1
13.128	32.872	0	23.542	17.473	1	31.752	14.248	0
13.584	11.642	1	23.801	22.199	0	32.020	13.980	0
14.990	16.355	1	24.115	16.195	1	33.789	12.211	0
16.048	2.297	1	24.176	2.509	1			

TABLE 3.3: Simulated restricted follow-up data ($n = 68$) used in the illustrative analysis: entry is time of entry to the trial (months), time is time to death or to the analysis (months); cens is censoring indicator (0=censored, 1=died).

Entry	Time	Cens	Entry	Time	Cens	Entry	Time	Cens
0.824	12.000	0	16.212	3.034	1	24.620	12.000	0
2.264	12.000	0	16.844	2.157	1	25.692	0.386	1
2.700	2.039	1	17.035	11.736	1	26.256	10.223	1
3.006	12.000	0	17.262	12.000	0	26.643	0.987	1
3.209	12.000	0	17.792	12.000	0	26.755	12.000	0
3.330	12.000	0	17.965	6.778	1	26.930	12.000	0
4.908	12.000	0	18.741	9.080	1	27.098	12.000	0
5.169	12.000	0	18.766	2.662	1	28.241	3.971	1
5.553	7.575	1	19.428	12.000	0	28.252	3.674	1
5.729	12.000	0	20.277	1.084	1	28.637	12.000	0
6.882	4.985	1	20.430	12.000	0	28.781	12.000	0
7.733	11.474	1	20.526	12.000	0	29.937	3.076	1
7.842	12.000	0	20.570	2.801	1	29.985	12.000	0
8.086	12.000	0	21.506	2.196	1	30.291	0.522	1
9.160	12.000	0	21.570	5.477	1	30.351	12.000	0
10.472	6.951	1	22.023	12.000	0	30.732	12.000	0
10.934	0.421	1	22.632	12.000	0	30.854	10.580	1
12.941	1.783	1	22.686	12.000	0	31.566	1.713	1
12.993	8.493	1	23.093	12.000	0	31.623	9.374	1
13.128	12.000	0	23.542	12.000	0	31.752	12.000	0
13.584	11.642	1	23.801	12.000	0	32.020	12.000	0
14.990	12.000	0	24.115	12.000	0	33.789	12.000	0
16.048	2.297	1	24.176	2.509	1			

longer than landmark time x ($t_f > x$). For example, for testing the hypothesis $H_0 : S(12) = 35\%$ vs. alternative $H_1 : S(12) = 50\%$ with type I error rate $\alpha = 0.05$ and power 90%, the required sample sizes are 80, 71, 69 by using $Z_1(12)$, $Z_2(12)$ and $Z_3(12)$ tests, respectively.

```
########## Sample Size for Testing Survival Prob. at Fixed Time #############
### shape is the shape parameter of underlying survival distribution; S0   ###
##$ and S1 are survival probaiblity at x under the null and alternative;    ###
### alpha and beta are the type I and type II errors and power=1-beta;      ###
### ta and tf are the accrual duration and follow-up time; dist is used to  ###
### specify the underlying survival distribution for the trial design.      ###
###########################################################################
Size=function(S0, S1, x, alpha, beta)
{
  Lam0=-log(S0); Lam1=-log(S1)
  sig2=1/S1-1
  z0=qnorm(1-alpha); z1=qnorm(1-beta)
  logit=function(t){log(t/(1-t))}
  n.Z1=(z0+z1)^2*sig2/((log(Lam0)-log(Lam1))^2*Lam1^2)
  v2=S1*sig2/(4*(1-S1))
  n.Z2=(z0+z1)^2*v2/(asin(sqrt(S1))-asin(sqrt(S0)))^2
  n.Z3=(z0+z1)^2*sig2/((logit(S1)-logit(S0))^2*(1-S1)^2)
  ans=c(n.Z1=ceiling(n.Z1),n.Z2=ceiling(n.Z2),n.Z3=ceiling(n.Z3))
  return(ans)
}
Size(S0=0.35,S1=0.50,x=12,alpha=0.1,beta=0.1)
n.Z1 n.Z2 n.Z3
  80   71   69
```

3.2.4 Comparison

We conducted simulation to study the performance of the proposed three test statistics. The parameter setting for the simulation is given as follows: the shape parameter κ is set to be $\kappa = 0.5, 1$ and 2; landmark time point $x = 1$ or 1.5; follow-up period $t_f = 1$; survival probability under null is set to $S_0(x) = 0.3$ to 0.6 by 0.1; the study is designed to detect a 20% increasing of the survival probability at landmark time point x for the alternative; uniform accrual with accrual period $t_a = 2$; type I error rate $\alpha = 0.05$ and power of 80%. For each design parameter configuration, sample size (n) is calculated for each test $Z_1(x)$, $Z_2(x)$ and $Z_3(x)$, and 10,000 random samples were generated from the Weibull distribution to estimate empirical type I error rate ($\hat{\alpha}$) and empirical power (EP) which were recorded in Table 3.4 for the case of $x > t_f$ ($x = 1.5$ and $t_f = 1$) and Table 3.5 for the case of $x = t_f(= 1)$. The simulation results showed that the log-log transformed test statistic $Z_1(x)$ can maintain the type I error close to the nominal level but the sample sizes were overestimated (empirical power was greater than the nominal power). The logit transformed test $Z_3(x)$ was liberal (type I error rate was always greater than the nominal level) and sample size was also overestimated particularly for a large value of $S_0(x)$. The arcsin-square root transformed test $Z_2(x)$ provided more accurate sample size estimation but it was slightly liberal. Furthermore,

TABLE 3.4: Sample sizes (n) were calculated under the Weibull distribution with nominal type I error $\alpha = 0.05$ and power of 80% for Nelson-Aalen estimate based tests. The corresponding empirical type I error ($\hat{\alpha}$) and empirical power (EP) were estimated based on 10,000 simulation runs for the case $x > t_f$ where $x = 1.5$ and follow-up time $t_f = 1$.

			Nelson-Aalen Estimate Based Tests						
			$Z_1(x)$			$Z_2(x)$			$Z_3(x)$
Shape	(S_0, S_1)	n	$\hat{\alpha}$	EP	n	$\hat{\alpha}$	EP	n	$\hat{\alpha}$	EP
0.5	(0.3, 0.5)	44	.051	.857	38	.057	.836	36	.056	.813
	(0.4, 0.6)	48	.048	.858	40	.059	.824	41	.061	.839
	(0.5, 0.7)	49	.048	.878	38	.066	.829	43	.060	.859
	(0.6, 0.8)	47	.047	.899	32	.051	.812	42	.051	.884
1	(0.3, 0.5)	45	.053	.850	39	.060	.827	37	.067	.824
	(0.4, 0.6)	50	.048	.864	41	.063	.828	42	.060	.836
	(0.5, 0.7)	50	.046	.872	39	.064	.831	44	.057	.856
	(0.6, 0.8)	48	.048	.899	34	.068	.834	43	.052	.880
2	(0.3, 0.5)	47	.053	.854	41	.062	.825	39	.066	.825
	(0.4, 0.6)	52	.050	.866	43	.063	.829	44	.060	.838
	(0.5, 0.7)	52	.048	.874	41	.064	.829	46	.057	.850
	(0.6, 0.8)	50	.045	.898	35	.067	.829	45	.053	.882

the arcsin-square root transformed test $Z_2(x)$ requires the smallest sample size among the three tests.

3.3   Kaplan-Meier Estimate Based Tests

Sample size calculation for evaluating survival probabilities using Kaplan-Meier based test statistics has been discussed by Nagashima et al. (2020). It is well known that the Kaplan-Meier estimate is asymptotically equivalent to the Nelson-Aalen estimate. Thus, in terms of large sample properties, the test statistics based on the Kaplan-Meier estimate are similar to the test statistics based on the Nelson-Aalen estimate. However, single-arm phase II trial design is often conducted with small sample size, therefore, it is worth exploring the performance of the three tests $Z_1(x)$, $Z_2(x)$ and $Z_3(x)$ based on the Kaplan-Meier estimate.

TABLE 3.5: Sample sizes (n) were calculated with nominal type I error $\alpha = 0.05$ and power of 80% for Nelson-Aalen estimate based tests. The corresponding empirical type I error ($\hat{\alpha}$) and empirical power (EP) were estimated based on 10,000 simulation runs under the exponential distribution for the case $x \leq t_f$, where $x = 1$ and follow-up time $t_f = 1$.

| | Nelson-Aalen Estimate Based Tests | | | | | | | |
| | $Z_1(x)$ | | | $Z_2(x)$ | | | $Z_3(x)$ | | |
(S_0, S_1)	n	$\hat{\alpha}$	EP	n	$\hat{\alpha}$	EP	n	$\hat{\alpha}$	EP
(0.3, 0.5)	43	.063	.886	37	.057	.837	35	.071	.847
(0.4, 0.6)	47	.044	.865	39	.055	.829	40	.068	.866
(0.5, 0.7)	48	.055	.900	37	.049	.807	42	.044	.834
(0.6, 0.8)	46	.038	.886	32	.059	.825	41	.062	.899

Note: The study design is distribution-free, therefore, simulation results is valid for any survival distribution.

3.3.1 Test Statistics

Consider the Kaplan-Meier estimate of the survival function $S(x) = P(T > x)$ which is given by

$$\hat{S}(x) = \prod_{u \leq x} \left\{ 1 - \frac{\Delta \bar{N}(u)}{\bar{Y}(u)} \right\},$$

where $\Delta \bar{N}(u) = \bar{N}(u) - \bar{N}(u^-)$ (a discrete version of $d\bar{N}(u)$). To derive the asymptotic distribution of the Kaplan-Meier estimate $\hat{S}(t)$, let $\tilde{\Lambda}(x)$ be the Nelson-Aalen estimate of cumulative hazard function $\Lambda(x)$, and since $\sqrt{n}\{\tilde{\Lambda}(x) - \Lambda(x)\} \xrightarrow{D} N(0, \sigma^2(x))$, applying the functional delta-method to $\tilde{S}(x) = e^{-\tilde{\Lambda}(x)}$, we get

$$\sqrt{n}\{\tilde{S}(x) - S(x)\} = \sqrt{n}\{g(\tilde{\Lambda}(x)) - g(\Lambda(x))\} \xrightarrow{D} N\{0, [g'(\Lambda(x))]^2 \sigma^2(x)\},$$

where $g(x) = e^{-x}$ and $g'(x) = -e^{-x}$ is the derivative of $g(x)$, and notation $\xrightarrow{D}$ denoted as 'converge in distribution'. Thus, $\sqrt{n}\{\tilde{S}(x) - S(x)\}$ converges in distribution to a zero-mean Gaussian process with variance function $S^2(x)\sigma^2(x)$. It can be shown that

$$\sqrt{n}\{\hat{S}(x) - \tilde{S}(x)\} \xrightarrow{P} 0,$$

where notation $\xrightarrow{P} 0$ denoted as 'converge in probability'. Thus, it follows from Slutsky's theorem, $\sqrt{n}\{\hat{S}(x) - S(x)\}$ converges in distribution to a zero-mean Gaussian process with asymptotic variance function

$$\text{Var}\{\sqrt{n}\hat{S}(x)\} \approx S^2(x)\sigma^2(x) = S^2(x) \int_0^x \frac{d\Lambda(u)}{\pi(u)},$$

and it can be estimated by

$$\widehat{\text{Var}}\{\sqrt{n}\hat{S}(x)\} = \hat{S}^2(x)\int_0^x \frac{d\bar{N}(u)}{\bar{Y}^2(u)/n} = \hat{S}^2(x)\hat{\sigma}^2(x).$$

which is asymptotically equivalent to Greenwood's formula.

Thus, all three tests $Z_1(x)$, $Z_2(x)$ and $Z_3(x)$ based on the Kaplan-Meier estimate have the same forms as given in equations (3.5), (3.6) and (3.7) and sample size formulae are also the same as given in equations (3.9), (3.10) and (3.11) for the $Z_1(x)$, $Z_2(x)$ and $Z_3(x)$, respectively.

3.3.2 Comparison

We conduct simulation to study the performance of three tests based on the Kaplan-Meier estimate. The parameter setting for the simulation is the same as in Section 3.2.4. The simulation results were recorded in Table 3.6 for the case of $x > t_f$ and Table 3.7 for the case of $x \leq t_f$. The results showed that the log-log transformed test $Z_1(x)$ preserved the type I error well but is slightly conservative and sample sizes were overestimated, particularly for a large survival probability $S_0(x)$. The arcsin-square root transformed test $Z_2(x)$ preserved the type I error well but was slightly liberal when $S_0(x) \geq 0.5$ and provided accurate sample size estimation. The logit transformed test $Z_3(x)$ was liberal and sample size was overestimated when $S_0(x) \geq 0.5$. The arcsin square root transformed $Z_2(x)$ test required the smallest sample size among the three tests. However, we also observed that the Kaplan-Meier estimate based test $Z_1(x)$ was more conservative than the Nelson-Aalen estimate based test and the Kaplan-Meier estimate based tests $Z_2(x)$ and $Z_3(x)$ preserved the type I error better than that of the Nelson-Aalen estimate based tests. The empirical powers of the Kaplan-Meier estimate based tests were closer to nominal level than the Nelson-Aalen estimate based tests, particularly for the test $Z_2(x)$. Therefore, in general, the Kaplan-Meier estimate based tests performed slightly better than the Nelson-Aalen estimate based tests in small sample sizes.

3.3.3 Study Design and Data Analysis

Suppose we design a single-arm phase II trial for patients with recurrent or refractory non-CNS rhabdoid tumors. The primary endpoint of the study is 1-year PFS. St. Jude Children's rhabdoid tumor study of 13 patients will be used as the literature data for the study design using Kaplan-Meier and Nelson-Aalen estimates based tests. The Kaplan-Meier (Nelson-Aalen) estimate of 1-year RFS was 46% (48%) which is treated as a reference value for the trial design (Figure 3.2). Investigators use molecular agent alisertib as new treatment and test the hypothesis $H_0 : S(1) \leq 0.46$ vs $H_1 : S(1) > 0.46$. The new treatment is promising if the 1-year PFS is at least 66%, that is 1-year survival probability at alternative $S(1) = 0.66$ or a 20% increasing in 1-year

TABLE 3.6: Sample sizes (n) were calculated under the Weibull distribution with nominal type I error rate $\alpha = 0.05$ and power of 80% for the Kaplan-Meier estimate based tests. The corresponding empirical type I error ($\hat{\alpha}$) and empirical power (EP) were estimated based on 10,000 simulation runs for the case of $x > t_f$ where $x = 1.5$ and follow-up time $t_f = 1$.

			Kaplan-Meier Estimate Based Tests							
			$Z_1(x)$			$Z_2(x)$			$Z_3(x)$	
Shape	(S_0, S_1)	n	$\hat{\alpha}$	EP	n	$\hat{\alpha}$	EP	n	$\hat{\alpha}$	EP
0.5	(0.3, 0.5)	44	.042	.834	38	.052	.817	36	.050	.802
	(0.4, 0.6)	48	.042	.845	40	.042	.799	41	.059	.820
	(0.5, 0.7)	49	.047	.860	38	.045	.786	43	.047	.832
	(0.6, 0.8)	47	.034	.878	32	.054	.806	42	.049	.879
1	(0.3, 0.5)	45	.043	.830	39	.050	.802	37	.051	.791
	(0.4, 0.6)	50	.042	.847	41	.052	.805	42	.050	.809
	(0.5, 0.7)	50	.041	.861	39	.053	.810	44	.050	.836
	(0.6, 0.8)	48	.041	.892	34	.055	.812	43	.048	.869
2	(0.3, 0.5)	47	.042	.832	41	.050	.808	39	.055	.802
	(0.4, 0.6)	52	.043	.846	43	.050	.802	44	.050	.810
	(0.5, 0.7)	52	.041	.856	41	.057	.809	46	.052	.838
	(0.6, 0.8)	50	.043	.891	35	.064	.819	45	.050	.874

TABLE 3.7: Sample sizes (n) were calculated with nominal type I error $\alpha = 0.05$ and power of 80% for Kaplan-Meier estimate based tests. The corresponding empirical type I error ($\hat{\alpha}$) and empirical power (EP) were estimated based on 10,000 simulation runs under the exponential distribution for the case of $x \leq t_f$, where $x = 1$ and follow-up time $t_f = 1$.

		Kaplan-Meier Estimate Based Tests								
		$Z_1(x)$			$Z_2(x)$			$Z_3(x)$		
(S_0, S_1)	n	$\hat{\alpha}$	EP	n	$\hat{\alpha}$	EP	n	$\hat{\alpha}$	EP	
(0.3, 0.5)	43	.030	.816	37	.057	.837	35	.035	.752	
(0.4, 0.6)	47	.044	.865	39	.055	.829	40	.037	.786	
(0.5, 0.7)	48	.029	.836	37	.049	.807	42	.044	.834	
(0.6, 0.8)	46	.038	.886	32	.059	.825	41	.062	.899	

Note: The study design is distribution-free, therefore, simulation results is valid for any survival distribution.

PFS. Assume uniform accrual with accrual duration $t_a = 2$ years and follow-up duration $t_f = 1$ year and no loss to follow-up. With a one-sided type I error rate 5% and power of 80%, using R code 'Size', the total sample sizes for the study are 48, 39 and 41 for the test statistics $Z_1(1)$, $Z_2(1)$ and $Z_3(1)$, respectively. In this design, the landmark time point $x = t_f = 1$, thus, as we have seen that the study design is free of the underlying survival distribution.

We generate a random sample from the Weibull distribution with shape parameter $\kappa = 1.32$ and scale parameter determined by the alternative $H_1 : S(1) = 0.66$ as the data to illustrate data analysis after the trial. For testing the null hypothesis $H_0 : S_0(1) = 0.46$, following R code 'Test' gives the Kaplan-Meier 1-year survival probability $\hat{S}(1) = 0.708$ (std.err $= 0.0656$) and p-value from Kaplan-Meier based tests are 0.0013, 0.0002 and 0.0005, respectively, for $Z_1(1)$, $Z_2(1)$ and $Z_3(1)$ for the random sample generated under the alternative hypothesis. Thus, we reject the null hypothesis and conclude that the new treatment is promising based on all three tests. The same conclusion can be made based on the Nelson-Aalen estimate based tests.

```
##### Sample size calculation for Testing Survival Prob. at Fixed Time   ######
### shape is the shape parameter of underlying survival distribution; S0    ###
##$ and S1 are survival probaiblity at x under the null and alternative;    ###
### alpha and beta are the type I and type II errors and power=1-beta;      ###
### ta and tf are the accrual duration and follow-up time; dist is used to  ###
### specify the underlying survival distribution for the trial design.      ###
################################################################################
Size=function(shape, x, S0, S1, alpha, beta, ta, tf, dist)
{    tau=ta+tf
     if (dist=="WB"){
     s=function(a,b,u){1-pweibull(u,a,b)}
     f=function(a,b,u){dweibull(u,a,b)}
     h=function(a,b,u){f(a,b,u)/s(a,b,u)}
     scale1=x/(-log(S1))^(1/shape)
     }

     if (dist=="LN"){
     s=function(a,b,u){1-plnorm(u,b,a)}
     f=function(a,b,u){dlnorm(u,b,a)}
     h=function(a,b,u){f(a,b,u)/s(a,b,u)}
     scale1=log(x)-shape*qnorm(1-S1)
     }

     if (dist=="LG"){
     s=function(a,b,u){1/(1+(u/b)^a)}
     f=function(a,b,u){(a/b)*(u/b)^(a-1)/(1+(u/b)^a)^2}
     h=function(a,b,u){f(a,b,u)/s(a,b,u)}
     scale1=x/(1/S1-1)^(1/shape)
     }

     if (dist=="GM"){
     s=function(a,b,u){1-pgamma(u,a,b)} ## shape=a; scale=b
     f=function(a,b,u){dgamma(u,a,b)}
     h=function(a,b,u){f(a,b,u)/s(a,b,u)}
     root1=function(t){s(shape,t,x)-S1}
     scale1=uniroot(root1,c(0,10))$root
```

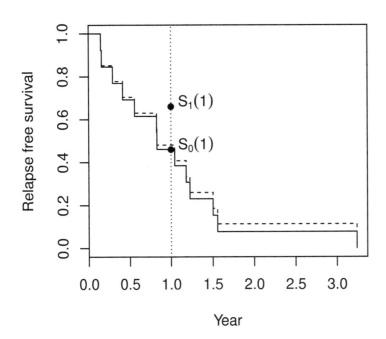

FIGURE 3.4: Kaplan-Meier survival curve (solid line) and Nelson-Aalen survival curve (dash line) and hypothesis values.

```
   }
 G=function(t){1-punif(t, tf, tau)}
 g=function(u){h(shape,scale1,u)/(s(shape,scale1,u)*G(u))}
 Lam0=-log(S0); Lam1=-log(S1)
 sig2=integrate(g, 0, x)$value
 z0=qnorm(1-alpha); z1=qnorm(1-beta)
 logit=function(t){log(t/(1-t))}
 n.Z1=(z0+z1)^2*sig2/((log(Lam0)-log(Lam1))^2*Lam1^2)
 v2=S1*sig2/(4*(1-S1))
 n.Z2=(z0+z1)^2*v2/(asin(sqrt(S1))-asin(sqrt(S0)))^2
 n.Z3=(z0+z1)^2*sig2/((logit(S1)-logit(S0))^2*(1-S1)^2)
 ans=c(n.Z1=ceiling(n.Z1),n.Z2=ceiling(n.Z2),n.Z3=ceiling(n.Z3))
 return(ans)
}
Size(shape=1.32,x=1,S0=0.46,S1=0.66,ta=2,tf=1,alpha=0.05,beta=0.2,dist="WB")
n.Z1 n.Z2 n.Z3
   48   38   41


######### Calculate Kaplan-Meier Based Test Statistic and p-value  ############
### shape is the Weibull shape parameter; S0 and S1 are survival probability###
### at time x under null and alternative; alpha and beta are the type I and ###
### type II errors; ta and tf are the accrual and follow-up time;          ###
### n is the total sample size.                                            ###
###############################################################################
library(survival)
Test=function(shape, x, S0, S1, ta, tf, n)
{scale1=x/(-log(S1))^(1/shape)
 tau=ta+tf
 logit=function(t){log(t/(1-t))}
 set.seed(534)
 t=rweibull(n, shape, scale1)
 u=runif(n, 0, ta)
 time=pmin(t, ta+tf-u)
 status=as.numeric(t<ta+tf-u)
 surv=Surv(time, status)
 ## Kaplan-Meier estimate using survfit ##
 fitKM<-survfit(surv~1)
 std.err.KM=tail(summary(fitKM)$std[summary(fitKM)$time<=x],1)
 S.hat.KM=tail(summary(fitKM)$surv[summary(fitKM)$time<=x],1)
 Lam0=-log(S0)
 Lam.hat.KM=-log(S.hat.KM)
 Z1.KM=(log(Lam0)-log(Lam.hat.KM))*Lam.hat.KM/(std.err.KM/S.hat.KM)
 nu.hat.KM=std.err.KM^2/(4*S.hat.KM*(1-S.hat.KM))
 Z2.KM=(asin(sqrt(S.hat.KM))-asin(sqrt(S0)))/sqrt(nu.hat.KM)
 Z3.KM=(logit(S.hat.KM)-logit(S0))*(1-S.hat.KM)/(std.err.KM/S.hat.KM)
 ##   Nelson-Aalen estimate using coxph ##
 fitNA=survfit(coxph(Surv(time,status)~1), type = "aalen")
 std.err.NA=tail(summary(fitNA)$std[summary(fitNA)$time<=x0],1)
 S.hat.NA=tail(summary(fitNA)$surv[summary(fitNA)$time<=x0],1)
 Lam.hat.NA=-log(S.hat.NA)
 Z1.NA=(log(Lam0)-log(Lam.hat.NA))*Lam.hat.NA/(std.err.NA/S.hat.NA)
 nu.hat.NA=std.err.NA^2/(4*S.hat.NA*(1-S.hat.NA))
 Z2.NA=(asin(sqrt(S.hat.NA))-asin(sqrt(S0)))/sqrt(nu.hat.NA)
 Z3.NA=(logit(S.hat.NA)-logit(S0))*(1-S.hat.NA)/(std.err.NA/S.hat.NA)

 p.Z1.KM=round(1-pnorm(Z1.KM),4)
```

```
p.Z2.KM=round(1-pnorm(Z2.KM),4)
p.Z3.KM=round(1-pnorm(Z3.KM),4)
ansKM=c(p.Z1.KM=p.Z1.KM,p.Z2.KM=p.Z2.KM, p.Z3.KM=p.Z3.KM,
    S.hat.KM=round(S.hat.KM,4),std.err.KM=round(std.err.KM,4))
p.Z1.NA=round(1-pnorm(Z1.NA),4)
p.Z2.NA=round(1-pnorm(Z2.NA),4)
p.Z3.NA=round(1-pnorm(Z3.NA),4)
ansNA=c(p.Z1.NA=p.Z1.NA,p.Z2.NA=p.Z2.NA, p.Z3.NA=p.Z3.NA,
    S.hat.NA=round(S.hat.NA,4),std.err.NA=round(std.err.NA,4))
return(c(ansKM, ansNA))
}
Test(shape=1.32,x=1,S0=0.46,S1=0.66,ta=2,tf=1,n=48)
p.Z1.KM   p.Z2.KM   p.Z3.KM   S.hat.KM std.err.KM
 0.0013    0.0002    0.0005    0.7083    0.0656
p.Z1.NA   p.Z2.NA   p.Z3.NA   S.hat.NA std.err.NA
 0.0011    0.0002    0.0004    0.7114    0.0651
```

3.4 Test Median Survival Time

To design a single-arm phase II trial with a survival endpoint, increasing median survival time is always clinically meaningful for the investigators. For example, median survival time of the standard of care is $m_0 = 6$ (months), investigators want to test the hypothesis: the new treatment can increase the median survival to $m_1 = 9$ months. Therefore the trial is designed by testing the following hypothesis

$$H_0 : m \le m_0 \quad vs \quad H_1 : m > m_0, \tag{3.15}$$

and powered at the alternative $m_1 = 9$ (months) (Figure 3.3).

3.4.1 Test Statistics

In the previous chapter, we have shown that increasing median survival time is equivalent to decreasing the hazard rate under the Weibull model. Testing the median survival time in the context of a nonparametric setting, is difficult because the nonparametric estimate of median survival time involves the density estimation which is often unrealistic for a small sample phase II trial (Wu, 2008, page 30). However, we can translate the median survival time into a parameter of the cumulative hazard function. Thus, we can use cumulative hazard based test statistics to test the hypothesis for the median survival time. Specifically, let the landmark time point $x = m_0$, the median survival time under the null. By noting $\Lambda_0(m_0) = -\log S_0(m_0) = \log(2)$, the log-log transformed test statistic for testing median survival time is given by

$$Z_1(m_0) = \frac{\sqrt{n}\{\log(\log 2) - \log \hat{\Lambda}(m_0)\}\hat{\Lambda}(m_0)}{\hat{\sigma}(m_0)},$$

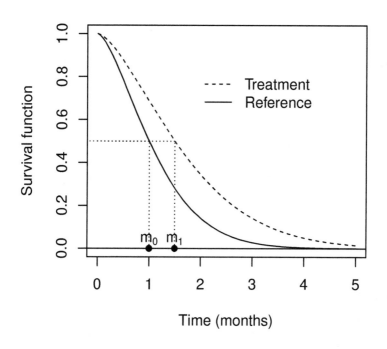

FIGURE 3.5: Testing median survival times.

where $\hat{\Lambda}(m_0)\}$ and $\hat{\sigma}^2(m_0)$ can be estimated based on either the Nelson-Aalen method or the Kaplan-Meier method.

3.4.2 Sample Size Formula

Using the same argument given in the previous section, sample size formula for testing median survival time using $Z_1(m_0)$ test is given by

$$n = \frac{(z_{1-\alpha} + z_{1-\beta})^2 \sigma_1^2(m_0)}{\{\log(\log 2) - \log \Lambda_1(m_0)\}^2 \Lambda_1^2(m_0)}, \tag{3.16}$$

where $\Lambda_1(x)$ is the cumulative hazard function under the alternative $m = m_1$. If we use arcsin-square root transformed test, it is given by

$$Z_2(m_0) = \frac{\sqrt{n}\{\arcsin\sqrt{\hat{S}(m_0)} - \arcsin\sqrt{0.5}\}}{\hat{\nu}(m_0)}, \tag{3.17}$$

where

$$\hat{\nu}^2(m_0) = \frac{\hat{S}(m_0)\hat{\sigma}^2(m_0)}{4\{1 - \hat{S}(m_0)\}},$$

and corresponding sample size formula is

$$n = \frac{(z_{1-\alpha} + z_{1-\beta})^2 \nu_1^2(m_0)}{\{\arcsin\sqrt{S_1(m_0)} - \arcsin\sqrt{0.5}\}^2}, \tag{3.18}$$

where

$$\nu_1^2(x) = \frac{S_1(m_0)\sigma_1^2(m_0)}{4\{1 - S_1(m_0)\}}.$$

The logit transformation test statistics is given by

$$Z_3(m_0) = \frac{\sqrt{n}\{\text{logit}\,\hat{S}(m_0) - \text{logit}(0.5)\}\{1 - \hat{S}(m_0)\}}{\hat{\sigma}(m_0)}. \tag{3.19}$$

and corresponding sample size formula is

$$n = \frac{(z_{1-\alpha} + z_{1-\beta})^2 \sigma_1^2(m_0)}{\{\text{logit}\,S_1(m_0) - \text{logit}(0.5)\}^2\{1 - S_1(m_0)\}^2}. \tag{3.20}$$

Remark 3: For testing median survival time $H_0 : m = m_0$, the maximum follow-up time should be longer than m_0 or each patient is followed for m_0 if we are interested in the median survival time only.

Remark 4: For testing the median survival time based on tests $Z_1(m_0)$, $Z_2(m_0)$ and $Z_3(m_0)$, sample size calculation depends on not only the asymptotic variance $\sigma_1^2(m_0)$ but also the value of survival distribution at the alternative at median survival time, that is $S_1(m_0)$ or $\Lambda_1(m_0)$. Thus, we have to make assumptions for the survival distribution at the alternative for the study design.

Example 3 *Continuation of Trial for Rhabdoid Tumors*

   Suppose that we wish to design a new trial and consider that the molecular agent alisertib is not worthy of further evaluation if the median RFS is at most 0.8 year and it is promising if the median RFS is at least 1.2 year. For the trial design, we assume that the RFS time follows the Weibull distribution $S(t) = e^{-\lambda t^{\kappa}}$, where $\lambda = \log(2)/m^{\kappa}$ and $\kappa = 1.32$; uniform accrual with accrual period $t_a = 2$ years; follow-up period $t_f = 1$ year; no loss to follow-up. With a 5% one-sided type I error rate and power of 80%, a total of sample size $n = 66$, 54 and 59 are required for the tests $Z_1(m_0)$, $Z_2(m_0)$ and $Z_3(m_0)$, respectively. The corresponding empirical powers using the Kaplan-Meier estimate based tests are 0.81, 0.84 and 0.85 for $Z_1(m_0)$, $Z_2(m_0)$ and $Z_3(m_0)$, respectively.

```
########## Sample Size Calculation for Testing Median Survival Time ###########
### shape is the Weibull shape parameter; m0 and m1 are median survival of  ###
### null and alternative; alpha and beta are the type I and type II errors  ###
### and power=1-beta; ta and tf are the accrual and follow-up time.         ###
###############################################################################
Size=function(shape, m0, m1, alpha, beta, ta, tf)
{ tau=ta+tf
  s=function(a,b,u){1-pweibull(u,a,b)}
  f=function(a,b,u){dweibull(u,a,b)}
  h=function(a,b,u){f(a,b,u)/s(a,b,u)}
  scale1=m1/(log(2))^(1/shape)
  G=function(t){1-punif(t, tf, tau)}
  g=function(u){h(shape,scale1,u)/(s(shape,scale1,u)*G(u))}
  S1=s(shape, scale1, m0)
  Lam1=-log(S1)
  logit=function(t){log(t/(1-t))}
  sig2=integrate(g, 0, m0)$value
  z0=qnorm(1-alpha); z1=qnorm(1-beta)
  n.Z1=(z0+z1)^2*sig2/((log(log(2))-log(Lam1))^2*Lam1^2)
  v2=S1*sig2/(4*(1-S1))
  n.Z2=(z0+z1)^2*v2/(asin(sqrt(S1))-asin(sqrt(0.5)))^2
  n.Z3=(z0+z1)^2*sig2/((logit(S1)-logit(0.5))^2*(1-S1)^2)
  ans=c(n.Z1=ceiling(n.Z1), n.Z2=ceiling(n.Z2), n.Z3=ceiling(n.Z3))
  return(ans)
}
Size(shape=1.32,m0=0.8,m1=1.2,alpha=0.05,beta=0.2,ta=2,tf=1)
n.Z1 n.Z2 n.Z3
  66   54   59
```

4

Two-Stage Design Evaluating Survival Probabilities

4.1 Introduction

For ethics reasons, it is often required that single-arm phase II trials should stop early if there is evidence that the experimental therapy has no anti-tumor activity at all to prevent patients being treated with an inferior treatment. Thus, single-arm phase II trials are often designed with an interim analysis for stopping futility, which is called a two-stage design. For binary endpoint, two-stage optimal design and minimax design proposed by Simon (1989) are commonly used for the oncology single-arm phase II trial designs to evaluate the efficacy of the trial with a binary primary endpoint, e.g., tumor response (responder vs. non-responder). However, these designs require suspension of accrual at interim analysis until the patient's follow-up is completed. Thus, they may not be applicable to the time-to-event endpoints which are often required in long-term follow-up. Furthermore, censoring becomes an issue during the interim analysis for the trials with time-to-event endpoints, as it is unclear how subjects without complete follow-up should be handled.

The most common time-to-event endpoint for the oncology trials is the survival probability at a fixed landmark time point. Case and Morgan (2003) and Huang et al. (2010) developed optimal two-stage designs that do not suspend accrual when the interim analysis is conducted. Huang and Thomas (2014) extended the results to three-stage design and implemented them in R package 'Optim2Design'. In this chapter, we will present the optimal two-stage design developed by Case and Morgan (2003) and Huang et al. (2010).

Suppose the primary endpoint of the single-arm phase II trial is the survival probability at a clinically meaningful landmark time point x. Let $S_0(x)$ be the survival probability at x which investigators are no longer interested in and $S(x)$ be the survival probability at x of the new treatment, which is unknown. Then, the trial design can be based on testing the following one-sided hypothesis:

$$H_0 : S(x) \leq S_0(x) \quad vs \quad H_1 : S(x) > S_0(x), \tag{4.1}$$

for a given landmark time point x or equivalent to test the following hypothesis

49

for the cumulative hazard function $\Lambda(x) = -\log S(x)$,

$$H_0 : \Lambda(x) \geq \Lambda_0(x) \quad vs \quad H_1 : \Lambda(x) < \Lambda_0(x), \tag{4.2}$$

at landmark time point x, where $\Lambda_0(x) = -\log S_0(x)$. The study is powered at alternative $H_1 : S(x) = S_1(x)$ or $H_1 : \Lambda(x) = \Lambda_1(x)$, where $S_1(x)$ and $\Lambda_1(x)$ are the survival distribution and cumulative hazard function at alternative hypothesis evaluated at the landmark time point x.

4.2   Two-Stage Design Based on Log-Log Test

4.2.1   Test Statistics

To develop an optimal two-stage design for testing the hypothesis of survival probability at a landmark time point x, we make the assumption that each patient is followed until failure (e.g., disease progression or death), or for x unit time, whichever occurs first. Assume patients were accrued uniformly on interval $[0, \mathrm{MDA}]$, where MDA is the maximum duration of accrual. For a two-stage trial, when the data are examined at calendar time $t \leq \mathrm{MTSL}$, where t is measured from the start of the study and $\mathrm{MTSL} = \mathrm{MDA} + x$ is the maximum total study duration, we observe the time to failure $X_i(t) = T_i \wedge x \wedge (t - A_i)^+$ and failure indicator $\Delta_i(t) = I(T_i \leq x \wedge (t - A_i)^+)$, where A_i is the entry time of the ith subject, $i = 1, \cdots, n$. Further define failure process $N_i(x, t) = \Delta_i(t)I(X_i(t) \leq x)$ and at-risk process $Y_i(x, t) = I(X_i(t) \geq x)$, $\bar{N}(x, t) = \sum_{i=1}^{n} N_i(x, t)$ and $\bar{Y}(x, t) = \sum_{i=1}^{n} Y_i(x, t)$. On the basis of the observed data $\{X_i(t), \Delta_i(t), i = 1, \cdots, n\}$, the Nelson-Aalen estimate for the cumulative hazard function $\Lambda(x) = \int_0^x \lambda(u) du$ at calendar time t is given by

$$\hat{\Lambda}(x; t) = \int_0^x \frac{\bar{N}(du, t)}{\bar{Y}(u, t)}; \quad t > x \tag{4.3}$$

where $\bar{N}(du, t)$ is defined as the derivative of $\bar{N}(u, t)$ respect to u. For a fixed t, it has been shown that the process

$$\sqrt{n}\{\Lambda_0(x) - \hat{\Lambda}(x; t)\}; \quad t > x$$

converges weakly to a Gaussian process $W(x; t)$ with independent increments and variance function

$$\sigma^2(x; t) = \int_0^x \frac{\lambda(u)}{\pi(u; t)} du, \tag{4.4}$$

where $\pi(u; t) = P(X_1(t) \geq u)$, and the asymptotic variance $\sigma^2(x; t)$ of $\sqrt{n}\hat{\Lambda}(x; t)$ can be estimated by

$$\hat{\sigma}^2(x; t) = \int_0^x \frac{\bar{N}(du, t)}{\bar{Y}^2(u, t)/n} \tag{4.5}$$

(Breslow and Crowley, 1974; Fleming and Harrington, 1991). Lin et al. (1996) further showed that the two-variable process $\sqrt{n}\{\Lambda_0(x) - \hat{\Lambda}(x;t)\};\quad x < t$, converges weakly to a zero-mean Gaussian process $W(x;t)$ with $\text{cov}\{W(x;t), W(y;s)\} = \sigma^2(x \vee y; s \wedge t)$, where variance $\sigma^2(\cdot;\cdot)$ is given by equation (4.4).

Lin et al. (1996) proposed to testing the hypothesis (4.1) using the following log-log transformed sequential test statistic

$$Z(x;t) = \frac{\sqrt{n}\{\log \Lambda_0(x) - \log \hat{\Lambda}(x;t)\}\hat{\Lambda}(x;t)}{\hat{\sigma}(x;t)}, \tag{4.6}$$

and showed that $Z(x;t)$ as a process in t can be approximated by a Gaussian process.

To calculate the asymptotic variance $\sigma^2(x;t)$ under the null and alternative hypotheses, assuming subjects were recruited with a uniform distribution over the maximum duration of accrual period MDA and followed for an additional period of length x, and the maximum total study duration MTSL = MDA+x. We further assume no loss to follow-up. Then, under the null hypothesis, we have $\pi(u;t) = P(T_1 \wedge x \wedge (t - A_1)^+ \geq u) = P(T_1 \geq u)P(A_1 < t - u) = S_0(u)P(A_1 < t - u), 0 < u \leq x$, where A_1 is uniformly distributed on interval $[0, \text{MDA}]$. Thus from equation (4.4), under the null hypothesis, we have

$$\sigma_0^2(x;t) = \int_0^x \frac{\lambda_0(u)}{S_0(u)P(A_1 < t - u)}du, \quad x < t \leq \text{MTSL} \tag{4.7}$$

where $\lambda_0(\cdot)$ and $S_0(\cdot)$ are the hazard and survival function under the null, respectively. Under the alternative hypothesis, we have

$$\sigma_1^2(x;t) = \int_0^x \frac{\lambda_1(u)}{S_1(u)P(A_1 < t - u)}du, \quad x < t \leq \text{MTSL} \tag{4.8}$$

where $\lambda_1(\cdot)$ and $S_1(\cdot)$ are the hazard and survival function under the alternative, respectively.

4.2.2 Optimal Two-Stage Design

We will use the following notations for a two-stage design:

- MDA: Maximum duration of accrual.

- MTSL: Maximum total study length which is MDA+x.

- r: Accrual rate.

- t_1: Time of the interim analysis at stage 1.

- t_2: Time from the interim analysis to the end of accrual, that is $t_2 = (\text{MDA} - t_1)^+$.

- n: Maximum sample size of the study.

- n_1: Sample size for stage 1.

- n_2: Sample size for stage 2.

- PS_0: Stopping probability at first stage under the null.

- ES: Expected sample size $= n_1 + (1 - PS_0)n_2$.

- ETSL: Expected total study length $= t_1 + (1 - PS_0)(\text{MTSL} - t_1)$.

Consider a phase II design with a single interim analysis at prespecified time t_1; at the interim analysis, the trial may be stopped for futility but not for efficacy. More specifically, the two-stage design is given as follows.

Stage 1: Accrue n_1 patients between time 0 and time t_1. Each patient is followed until failure, or for x unit time, or until time t_1, whichever occurs first. More specifically, at stage 1, the observed failure time is $X_i(t_1) = T_i \wedge x \wedge (t_1 - A_i)^+$ and the observed failure indicator is $\Delta_i(t_1) = I(T_i \leq x \wedge (t_1 - A_i)^+)$, $i = 1, \cdots, n_1$, where A_i is the entry time of the ith subject. The test statistic $Z_1 = Z(x; t_1)$ is calculated and compared to the boundary value c_1. If $Z_1 < c_1$, the trial is stopped for futility, with stopping probability $PS_0 = \Phi(c_1)$ under the null hypothesis. Otherwise, the trial continues to stage 2.

Stage 2: Accrue n_2 additional patients between times t_1 and MDA. Each patient is followed until failure or for x unit time. At stage 2, the observed failure time is $X_i(\text{MTSL}) = T_i \wedge x \wedge (\text{MTSL} - Y_i)^+$ and the observed failure indicator is $\Delta_i(\text{MTSL}) = I(T_i \leq x \wedge (\text{MTSL} - A_i)^+)$, $i = 1, \cdots, n$. We calculate a second test statistic $Z_2 = Z(x; \text{MTSL})$ using all data (including stage 1 data), and we reject null hypothesis if $Z_2 > c_2$. The stopping boundary c_2 will also be derived as part of the optimal design. Figure 4.1 illustrates the survival data from a hypothetical single-arm two-stage phase II trial.

Overall type I error and power for the two-stage design satisfy the following two equations:

$$P(Z_1 > c_1, Z_2 > c_2 | H_0) = \alpha$$
$$P(Z_1 > c_1, Z_2 > c_2 | H_1) = 1 - \beta$$

Under the null hypothesis H_0, (Z_1, Z_2) is approximately bivariate normal distributed with mean zero and variance-covariance matrix

$$\Sigma_0 = \begin{pmatrix} 1 & \rho_0 \\ \rho_0 & 1 \end{pmatrix}$$

where correlation $\rho_0 = \sigma_{20}/\sigma_{10}$. The variances σ_{20}^2 and σ_{10}^2 can be calculated as follows:

$$\sigma_{10}^2 = \int_0^x \frac{\lambda_0(u)du}{S_0(u)P(A_1 \leq t_1 - u)} = \int_0^x \frac{\lambda_0(u)du}{S_0(u)G(u, t_1)} \qquad (4.9)$$

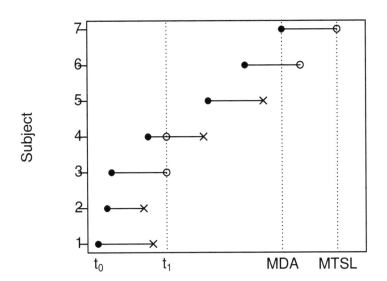

FIGURE 4.1: Survival data for two-stage trial with a restricted follow-up time x.

where $G(u, t_1) = P(A_1 \leq t_1 - u)$ with A_1 uniformly distributed on $[0, \text{MDA}]$, or

$$G(u, t_1) = \begin{cases} 1 & u \leq t_1 - \text{MDA} \\ (t_1 - u)/\text{MDA} & t_1 - \text{MDA} < u \leq t_1 \\ 0 & u > t_1 \end{cases}$$

and

$$\sigma_{20}^2 = \int_0^x \frac{\lambda_0(u)du}{S_0(u)P(A_1 \leq \text{MTSL} - u)} = \frac{1}{S_0(x)} - 1 \qquad (4.10)$$

by noting $P(A_1 \leq \text{MTSL} - u) = 1$ as $\text{MTSL-u} \geq \text{MTSL} - x = \text{MDA}$ for all $0 < u \leq x$.

Under alternatives H_1, (Z_1, Z_2) is approximately bivariate normal distributed with mean $\mu = (\rho_1 u, u)$ and variance-covariance matrix

$$\Sigma_1 = \begin{pmatrix} 1 & \rho_1 \\ \rho_1 & 1 \end{pmatrix}$$

where

$$u = \frac{\sqrt{n}\{\log \Lambda_0(x) - \log \Lambda_1(x)\}\Lambda_1(x)}{\sigma_{21}} \qquad (4.11)$$

and $\rho_1 = \sigma_{21}/\sigma_{11}$ with

$$\sigma_{11}^2 = \int_0^x \frac{\lambda_1(u)du}{S_1(u)P(A_1 \leq t_1 - u)} = \int_0^x \frac{\lambda_1(u)du}{S_1(u)G(u, t_1)} \qquad (4.12)$$

and

$$\sigma_{21}^2 = \int_0^x \frac{\lambda_1(u)du}{S_1(u)P(A_1 \leq \text{MTSL} - u)} = \frac{1}{S_1(x)} - 1, \qquad (4.13)$$

where $S_1(x)$ is the survival probability at a landmark time point under the alternative and $\Lambda_1(x) = -\log S_1(x)$.

4.2.3   Optimization Algorithm

The proposed two-stage design was implemented with four parametric distributions: Weibull, log-normal, gamma and log-logistic. In the design, we assumed that the distributions under null and alternative have a same shape parameter and a different scale parameter. An algorithm is specified to minimize the expected sample size (ES) as described in the following 11 steps. The user inputs are:

- The accrual rate r, assumed a uniform accrual during the accrual period.

- The type I error rate α and type II error rate β (or power $1 - \beta$).

- The survival probabilities under null $S_0(x)$ and under alternative $S_1(x)$ at a landmark time point x. The method is implemented with four parametric distributions: Weibull (WB), log-normal (LN), log-logistic (LG) and gamma (GM).

- The landmark time point x.

The 11 steps of the optimization method given by Huang et al. (2010) are listed as follows:

1. Fix initial values for n and ρ_1, combined with the type I error rate and power constraints, these determine all of the remaining parameters as described by steps 2-10.

2. Calculate MDA $= n/r$ for the projected accrual duration.

3. Calculate σ_{21} (H_1) from equation (4.13), and σ_{20} (H_0) from equation (4.10).

4. Obtain σ_{11} from $\rho_1 = \sigma_{21}/\sigma_{11}$, which implies $\sigma_{11} = \sigma_{21}/\rho_1$.

5. Obtain t_1 by solving equation (4.12) for the value of σ_{11} computed in the previous step. A unique root exists because σ_{11} is monotone in t_1.

6. Calculate $t_2 = (\text{MDA} - t_1)^+$.

7. Calculate σ_{10} from equation (4.9).

8. Calculate $\rho_0 = \sigma_{20}/\sigma_{10}$.

9. Calculate u from equation (4.11).

10. Obtain c_1 and c_2 by solving the following two equations:

$$B(c_1, c_2, \rho_0) - \alpha = 0 \qquad (A)$$
$$B(c_1 - \rho_1 u, c_2 - u, \rho_1) - (1 - \beta) = 0 \qquad (B)$$

where $B(c_1, c_2, \rho)$ denotes the probability of exceeding c_1 and c_2 for a standard bivariate normal distribution with correlation ρ. The two-variable equations are solved using R function 'optim' by minimizing $(A)^2 + (B)^2$ to 0.

11. The optimal design to minimize the ES $= n_1 + (1 - \text{PS})n_2$ is found by evaluating all possible n. The optimal ρ_1 for each n is found using a combination golden-section search and parabolic interpolation minimization as described by Brent (1973). It is implemented by the function 'optimize' in R. The range of ρ_1 is between 0 and 1. The search for optimal n can begin with the fixed sample size of the single-stage design. The two-stage design requires a larger maximum sample size than a single stage design and thus, the fixed sample size n_0 is a lower bound.

Remark 1: In the two-stage design, we assumed that the shape parameters of four distributions (Weibull, gamma, log-normal and log-logistic) are known (estimated from historical data) and the scale parameters of the corresponding distributions are calculated from $S_0(x)$ and $S_1(x)$ for the null and alternative survival distributions, respectively. For example, for the Weibull distribution $S(x) = e^{-(x/b)^a}$, where a and b are the shape and scale parameters and the scale parameters are solved to be $b_0 = x/\{-\log S_0(x)\}^{1/a}$ and $b_1 = x/\{-\log S_1(x)\}^{1/a}$ for the null and alternative survival distributions, respectively. Therefore, in general the distribution under the alternative does not satisfy the proportional hazard assumption to the null distribution (except for the exponential distribution).

Remark 2: If investigators feel that a proportional hazard model should be the satisfied model for the study design, then, the alternative survival distribution is $S_1(u) = [S_0(u)]^\delta$, where δ is the hazards ratio. The proposed two-stage design can be applied to this case with some simple modifications of the corresponding R code.

Remark 3: In the two-stage design, the first stage interim analysis time t_1 is matched to the first stage sample size n_1 by accrual rate r, that is $n_1 = rt_1$. However, in a real trial, they may not occur at the same time. To insure the power of interim analysis, we suggest that the interim analysis is performed at calendar time t_1 or after n_1 patients enrolled in the study, whichever occurs later.

Remark 4: In the case of a patient's follow-up time being longer than x unit time, because the test statistics use the data only up to x unit time, thus, study design and data analysis are the same as with the follow-up time restricted to x unit time.

4.2.4 Two-Stage Trial Data

To illustrate two-stage data for a single-arm phase II trial with restricted $x = 12$ months follow-up, stage I data is created at calendar time $t_1 = 20.430$ (months) with $n_1 = 34$ subjects using data recorded in Table 4.1. The observed survival times and censoring indicators of the stage I data are $X_i(t_1) = T_i \wedge x \wedge (t_1 - A_i)^+$ and $\Delta_i(t_1) = I(T_i \leq x \wedge (t_1 - A_i)^+), i = 1, \cdots, n_1$, where A_i is the entry time of ith subject which is uniformly distributed on interval $[0, 34]$ (months) and $t_1 = 20.43$ is the interim analysis time. The Nelson-Aalen estimate $\hat{S}(x)$ for stage I data is $\hat{S}(12) = 0.6$ and standard error of $\hat{S}(12)$ is std.err=0.1033. The Nelson-Aalen estimate of survival curve based on stage I data is given in Figure 4.2. Using $\Lambda_0(12) = -\log(0.35)$ and $\hat{\Lambda}(12) = -\log\{\hat{S}(12)\} = -\log(0.6)$, the observed value of $Z_1(x; t_1)$ test for stage I data

TABLE 4.1: Simulated restricted follow-up ($x = 12$ months) stage I data ($n_1 = 34$) used in the illustrative analysis. Entry: time of entry to the trial (months); Time: time to death or to the analysis (months); Cens: censoring indicator (0=censored, 1=died).

Entry	Time	Cens	Entry	Time	Cens	Entry	Time	Cens
0.824	12.000	0	7.842	12.000	0	16.844	2.157	1
2.264	12.000	0	8.086	12.000	0	17.035	3.395	0
2.700	2.039	1	9.160	11.270	0	17.262	3.168	0
3.006	12.000	0	10.472	6.951	1	17.792	2.638	0
3.209	12.000	0	10.934	0.421	1	17.965	2.465	0
3.330	12.000	0	12.941	1.783	1	18.741	1.689	0
4.908	12.000	0	12.993	7.437	0	18.766	1.664	0
5.169	12.000	0	13.128	7.302	0	19.428	1.002	0
5.553	7.575	1	13.584	6.846	0	20.277	0.152	0
5.729	12.000	0	14.990	5.440	0	20.430	0.000	0
6.882	4.985	1	16.048	2.297	1			
7.733	11.474	1	16.212	3.034	1			

is

$$Z_1(x;t_1) = \frac{\{\log \Lambda_0(12) - \log \hat{\Lambda}(12;t_1)\}\hat{\Lambda}(12;t_1)}{\text{std.err}\{\hat{S}(12;t_1)\}/\hat{S}(12;t_1)} = 2.1377 \qquad (4.14)$$

and the observed value of the test for stage II data (final data) is $Z_2(x, \text{MTSL}) = 3.3086$. Following R code 'Test' can be used to calculate the two-stage test statistics.

Remark: By noting $\hat{\sigma}(x;t) = \sqrt{n} \times \text{std.err}(\hat{\Lambda}(x;t))$ and $\text{std.err}\{\hat{\Lambda}(x;t) \simeq \text{std.err}\{\hat{S}(x;t)\}/\hat{S}(x;t)$, we have $\hat{\sigma}(x;t) = \sqrt{n} \times \text{std.err}\{\hat{S}(x;t)\}/\hat{S}(x;t)$ which leads to the equation (4.14)

```
########## Calculate Nelson-Aalen Estimate and Test Statistics  #############
### S0 is the survival probability at landmark time point x under the null; ###
### x is also the restricted follow-up time; t1 is interim analysis time;   ###
### n1 is stage I sample size; data is the data frame with variable Time:   ###
### time to death or the analysis and variable Cens: censoring indicator.   ###
### Entry: patient entry time.                                              ###
#############################################################################
library(survival)
Test=function(S0,x,t1,n1,data){
 Time=data$Time; Cens=data$Cens
 A=data$Entry
 fitNA=survfit(coxph(Surv(Time,Cens)~1), type = "aalen")
 std.err.NA=tail(summary(fitNA)$std[summary(fitNA)$time<=x],1)
 S.hat.NA=tail(summary(fitNA)$surv[summary(fitNA)$time<=x],1)
 Lam0=-log(S0); Lam.hat.NA=-log(S.hat.NA)
 Z2.MTSL=(log(Lam0)-log(Lam.hat.NA))*Lam.hat.NA/(std.err.NA/S.hat.NA)
```

```
nu.hat.NA=std.err.NA^2/(4*S.hat.NA*(1-S.hat.NA))
Z2.MTSL=round(Z2.MTSL,4)
Time.t1=pmin(Time,x,(t1-A))[1:n1]
Cens.t1=as.numeric(I(Time<pmin(x,(t1-A))))[1:n1]
fitNA.t1=survfit(coxph(Surv(Time.t1,Cens.t1)~1), type = "aalen")
std.err.NA.t1=tail(summary(fitNA.t1)$std[summary(fitNA.t1)$time<=x],1)
S.hat.NA.t1=tail(summary(fitNA.t1)$surv[summary(fitNA.t1)$time<=x],1)
Lam.hat.NA.t1=-log(S.hat.NA.t1)
Z1.t1=(log(Lam0)-log(Lam.hat.NA.t1))*Lam.hat.NA.t1/(std.err.NA.t1/S.hat.NA.t1)
Z1.t1=round(Z1.t1,4)
ans=c(Z1.t1=Z1.t1, S.hat.NA.t1=round(S.hat.NA.t1,4),
std.err.NA.t1=round(std.err.NA.t1,4),Z2.MTSL=Z2.MTSL,
S.hat.NA=round(S.hat.NA,4),std.err.NA=round(std.err.NA,4))
return(ans)
}
Test(S0=0.35,x=12,t1=20.430, n1=34, data)
  Z1.t1    S.hat.NA.t1  std.err.NA.t1   Z2.MTSL    S.hat.NA     std.err.NA
  2.1377       0.6000       0.1033      3.0458      0.5475       0.0601
```

4.2.5   Simulation

To study characteristics of the proposed optimal two-stage design, we conducted simulation under various scenarios. In the simulation, the survival distribution was taken as the Weibull distribution with shape parameter to be 0.5, 1 and 2; the survival probability under null $S_0(1)$ is set to be 0.3 to 0.6 by 0.1; the scale parameters under the null are determined by the value of $S_0(1)$; the survival probability under alternative $S_1(1)$ is set to be 0.5 to 0.8 by 0.1, or a 20% increase in survival probability at $x = 1$; accrual rate is set to be $r = 10$; type I error rate $\alpha = 0.05$ and power of 80%. Under each scenario, the R code 'TwoStage.CM' was used to create the two-stage design. Based on the output parameters of each two-stage design, we conducted a simulation study to estimate the empirical type I error and empirical power. The simulation results for the Nelson-Aalen estimate based test and Kaplan-Meier estimate based test were presented in Table 4.2. From simulations we have the following observations: The two-stage design for the Nelson-Aalen based test didn't preserve the type I error and power well. The empirical type I error could be inflated and the study could be overpowered. The two-stage design for the Kaplan-Meier based test did preserve the type I error well. However the study could be overpowered particularly for a relatively large survival probability under the null hypothesis. The empirical type I error could be inflated and study could be overpowered. The two-stage designs were very robust across the different shape parameters of the Weibull distribution.

Examples input and output using R code 'TwoStage.CM' are given as follows:

```
########## input parameters ####################
alpha=0.05
beta=0.2
r=10
S0=0.3
```

```
S1=0.5
shape=0.5
x=1
###### output for Weibull distribution #######
TwoStage.CM(shape,S0,S1,x,r,alpha,beta,dist="WB")
$test
alpha  beta shape    S0    S1    x    n0
 0.05  0.20  0.50  0.30  0.50  1.00 43.00

$result
     ES    ETSL
30.6275  3.3968

$StopProb
StopProbNull  StopProbAlt
      0.6660       0.1313

$n
 Stage1_n1 FinalMax_n
        21         50

$boundary
    C1     C2
0.4288 1.5254

$stageTime
    t1    t2  MTSL   MDA
2.091 2.909 6.000 5.000
```

4.2.6 Study Design and Data Analysis

Case and Morgan (2003) illustrated a single-arm phase II trial design to access chemo-radiation treatment for patients with resectable pancreatic cancer. The 12-month survival probability of 35% or less on treatment is considered inefficient, while 12-month survival probability of 50% or greater would be worthwhile. To illustrate two-stage trial design, assume uniform accrual with accrual rate 2 subjects per month and 12 months follow-up; survival times of experimental group follow an exponential distribution $S(t) = e^{-\lambda t}$ with parameter $\lambda = -\log S(12)/12 = -\log(0.5)/12$. Using optimal two-stage design R code 'TwoStage.CM', we obtain the following results:

```
TwoStage.CM(shape=1,S0=0.35,S1=0.5,x=12,r=2,alpha=0.05,beta=0.2,dist="WB")
$test
alpha  beta shape    S0    S1     x    n0
 0.05  0.20  1.00  0.35  0.50 12.00 75.00


$result
     ES    ETSL
58.6489 33.8067


$StopProb
StopProb.Null  StopProb.Alt
       0.6265        0.1185


$n
 Stage1.n1 FinalMax.n
        43         86


$boundary
      C1     C2
 0.3225 1.5458


$stageTime
      t1      t2    MTSL     MDA
21.1708 21.8292 55.0000 43.0000
```

This two-stage design can be executed as follows: At stage I, we will enroll 43 patients and the interim analysis will be performed after the 43rd patient enrolled in the study or at calendar time $t_1 = 21.1708$ (months), whichever occurs later. The stage I test statistic $Z_1(x; t_1)$ will be calculated and compared with stage 1 boundary $c_1 = 0.3225$. If $Z_1(x; t_1) < c_1$, the trial will stop for futility, otherwise, the trial will go to second stage. At stage II, an additional 43 patients will be enrolled in the study for a total of 86 patients. Once the final patient is enrolled in the study and followed for $x = 1$ year, the final analysis will be conducted to calculate the final test statistic $Z_2(x; \text{MTSL})$. If $Z_2(x; \text{MTSL}) < c_2 = 1.5458$, we will conclude that the new treatment is not worth further investigation. If $Z_2(x; \text{MTSL}) > c_2$, we will conclude that the new treatment is promising. The operating characteristics of this two-stage design is summarized in Table 6.3. The expected sample size for the two-stage

TABLE 4.2: Optimal two-stage characteristics for designs based on log-log transformed test under Weibull distribution. Overall empirical type I error rate ($\hat{\alpha}$) and empirical power (EP) for the two-stage designs were estimated from 10,000 simulated trials.

Nelson-Aalen Estimate Based Tests

Shape	(S_0, S_1)	t_1	c_1	c_2	ES	ETSL	n_1	n_{max}	PS_0	PS_1	$\hat{\alpha}$	EP
0.5	(.3, .5)	2.09	.429	1.525	30.6	3.40	21	50	.67	.13	.068	.87
	(.4, .6)	2.24	.439	1.525	33.2	3.65	23	55	.67	.13	.059	.87
	(.5, .7)	2.27	.440	1.526	33.7	3.70	23	56	.67	.13	.054	.89
	(.6, .8)	2.18	.444	1.524	32.4	3.56	22	54	.67	.13	.062	.91
1	(.3, .5)	2.20	.366	1.539	31.6	3.52	22	49	.64	.12	.060	.87
	(.4, .6)	2.36	.387	1.537	34.2	3.77	24	54	.65	.13	.042	.86
	(.5, .7)	2.37	.410	1.528	34.7	3.81	24	56	.65	.13	.055	.89
	(.6, .8)	2.29	.391	1.536	33.4	3.69	23	53	.65	.13	.045	.90
2	(.3, .5)	2.29	.339	1.542	32.5	3.62	23	49	.63	.12	.061	.87
	(.4, .6)	2.44	.357	1.540	35.1	3.87	25	54	.64	.12	.043	.87
	(.5, .7)	2.48	.359	1.540	35.6	3.92	25	55	.64	.12	.047	.88
	(.6, .8)	2.39	.363	1.538	34.3	3.79	24	53	.64	.12	.045	.90

Kaplan-Meier Estimate Based Tests

Shape	(S_0, S_1)	t_1	c_1	c_2	ES	ETSL	n_1	n_{max}	PS_0	PS_1	$\hat{\alpha}$	EP
0.5	(.3, .5)	2.09	.429	1.525	30.6	3.40	21	50	.67	.13	.037	.81
	(.4, .6)	2.24	.439	1.525	33.2	3.65	23	55	.67	.13	.056	.85
	(.5, .7)	2.27	.440	1.526	33.7	3.70	23	56	.67	.13	.031	.85
	(.6, .8)	2.18	.444	1.524	32.4	3.56	22	54	.67	.13	.034	.88
1	(.3, .5)	2.20	.366	1.539	31.6	3.52	22	49	.64	.12	.035	.80
	(.4, .6)	2.36	.387	1.537	34.2	3.77	24	54	.65	.13	.040	.84
	(.5, .7)	2.37	.410	1.528	34.7	3.81	24	56	.65	.13	.032	.85
	(.6, .8)	2.29	.391	1.536	33.4	3.69	23	53	.65	.13	.044	.89
2	(.3, .5)	2.29	.339	1.542	32.5	3.62	23	49	.63	.12	.036	.81
	(.4, .6)	2.44	.357	1.540	35.1	3.87	25	54	.64	.12	.040	.85
	(.5, .7)	2.48	.359	1.540	35.6	3.92	25	55	.64	.12	.045	.87
	(.6, .8)	2.39	.363	1.538	34.3	3.79	24	53	.64	.12	.044	.89

TABLE 4.3: The operating characteristics of the two-stage design.

Stage	n	Boundary	Interim time	ES	ETSL	SP_0	SP_1
1	43	0.3225	21.1708				
2	86	1.5458	55	58.6489	33.8067	0.6265	0.1185

design is 59 and expected total study length is 33.8 months, with the first stage stopping probabilities at null and alternative are 0.6265 and 0.1185, respectively.

To illustrate data analysis after a two-stage trial, a random sample of 86 survival times of experimental subjects were simulated from an exponential distribution $S(t) = e^{-\lambda t}$ with parameter $\lambda = -\log S(12)/12 = -\log(0.5)/12$. Assume patients were uniformly recruited at a rate of 2 patients per month with an accrual period of 43 months. There was no loss to follow-up, thus, censoring was administrative censoring only. The restricted follow-up full data and stage I data (with restricted follow-up time $x = 12$ months) were given in Tables 4.4 and 4.5. To calculate the test statistics, we have to calculate the Nelson-Aalen estimate $\hat{\Lambda}(x; t)$ and its variance estimate $\hat{\sigma}^2(x; t)$ at each stage. Applying R function 'coxph' to the stage I data, we obtain a Nelson-Aalen estimate of 12-month survival probability $\hat{S}(12; t_1) = 0.4175$ and standard error std.err$\{\hat{S}(12, t_1)\} = 0.0939$. The Nelson-Aalen estimate of survival curve is given in Figure 4.2. Using $\Lambda_0(12) = -\log(0.35)$ and $\hat{\Lambda}(12; t_1) = -\log \hat{S}(12; t_1) = -\log(0.4175)$, the observed value of log-log transformed stage I test statistic $Z_1(12; t_1)$ is obtained below

$$Z_1(12; t_1) = \frac{\{\log \Lambda_0(12) - \log \hat{\Lambda}(12; t_1)\}\hat{\Lambda}(12; t_1)}{\text{std.err}\{\hat{S}(12, t_1)\}/\hat{S}(12; t_1)} = 0.7137.$$

As $Z_1(12; t_1) > c_1 = 0.3225$, the trial goes to stage II to enroll 43 more patients. After stage II data were collected and analyzed, we have $\hat{S}(12; \text{MTSL}) = 0.4686$ and standard error of $\hat{S}(12; \text{MTSL})$ is std.err$\{\hat{S}(12, \text{MTSL})\} = 0.0536$. Thus, the observed value of stage II test statistic is given by

$$Z_2(12; \text{MTSL}) = \frac{\{\log(\Lambda_0(12)) - \log(\hat{\Lambda}(12; \text{MTSL}))\}\hat{\Lambda}(12; \text{MTSL})}{\text{std.err}\{\hat{S}(12, \text{MTSL})\}/\hat{S}(12; \text{MTSL})} = 2.1572.$$

As $Z_2(12; \text{MTSL}) > c_2 = 1.5458$, we reject the null hypothesis and conclude the experimental treatment is efficient. The following R code 'Test' is given for calculation of the two-stage test statistics and the corresponding standard errors.

```
########### Calculate Nelson-Aalen Estimate and Test Statistics ##############
### S0 is the survival probability at landmark time point x under the null; ###
```

```
### x is also the restricted follow-up time; t1 is interim analysis time;   ###
### n1 is stage I sample size; data is the data frame with variable Time:    ###
### time to death or the analysis and variable Cens: censoring indicator.    ###
### Entry: patient entry time.                                               ###
################################################################################
library(survival)
Test=function(S0,x,t1, n1, data){
 Time=data$Time; Cens=data$Cens
 A=data$Entry
 fitNA=survfit(coxph(Surv(Time,Cens)~1), type = "aalen")
 std.err.NA=tail(summary(fitNA)$std[summary(fitNA)$time<=x],1)
 S.hat.NA=tail(summary(fitNA)$surv[summary(fitNA)$time<=x],1)
 Lam0=-log(S0); Lam.hat.NA=-log(S.hat.NA)
 Z2.MTSL=(log(Lam0)-log(Lam.hat.NA))*Lam.hat.NA/(std.err.NA/S.hat.NA)
 Z2.MTSL=round(Z2.MTSL,4)
 Time.t1=pmin(Time,x,(t1-A))[1:n1]
 Cens.t1=as.numeric(I(Time<pmin(x,(t1-A))))[1:n1]
 fitNA.t1=survfit(coxph(Surv(Time.t1,Cens.t1)~1), type = "aalen")
 std.err.NA.t1=tail(summary(fitNA.t1)$std[summary(fitNA.t1)$time<=x],1)
 S.hat.NA.t1=tail(summary(fitNA.t1)$surv[summary(fitNA.t1)$time<=x],1)
 Lam.hat.NA.t1=-log(S.hat.NA.t1)
 Z1.t1=(log(Lam0)-log(Lam.hat.NA.t1))*Lam.hat.NA.t1/(std.err.NA.t1/S.hat.NA.t1)
 Z1.t1=round(Z1.t1,4)
 ans=c(Z1.t1=Z1.t1, S.hat.NA.t1=round(S.hat.NA.t1,4),
 std.err.NA.t1=round(std.err.NA.t1,4),Z2.MTSL=Z2.MTSL,
 S.hat.NA=round(S.hat.NA,4),std.err.NA=round(std.err.NA,4))
 return(ans)
}
Test(S0=0.35,x=12,t1=24.81,n1=43,data=data)
  Z1.t1   S.hat.NA.t1 std.err.NA.t1    Z2.MTSL    S.hat.NA   std.err.NA
 0.7137     0.4175        0.0939        2.1572      0.4686     0.0536
```

Once a two-stage designed trial has been completed, the p-value calculation should account for the two-stage nature of the procedure. If $Z_1(x; t_1) < c_1$, the trial will stop for futility, thus, p-value based on two-stage design can be calculated as follows:

$$p = P(Z_1(x; t_1) > z_1(x; t_1)) = 1 - \Phi(z_1(x; t_1)),$$

where $z_1(x; t_1)$ is the observed value of stage I test statistic $Z_1(x; t_1)$. If $Z_1(x; t_1) > c_1$, the trial will go to stage II, thus, p-value based on two-stage design can be calculated as follows:

$$\begin{aligned} p &= P(Z_1(x; t_1) > c_1, Z_2(x; \text{MTSL}) > z_2(x; \text{MTSL})) \\ &= B(c_1, z_2(x; \text{MTSL}), \rho), \end{aligned}$$

where $B(c_1, c_2, \rho)$ denotes the probability of exceeding c_1 and c_2 for a standard bivariate normal distribution with correlation $\rho = \hat{\sigma}(x, \text{MTSL})/\hat{\sigma}(x; t_1)$.

For the above example, $c_1 = 0.3225$, $Z_2(12; \text{MTSL}) = 2.1572$, $\rho = 0.5086$, and the trial goes to stage II, thus, p-value can be calculated as follows:

$$p = B(c_1, z_2(x; \text{MTSL}), \rho) = B(0.3225, 2.1572, 0.5086) = 0.01336,$$

which confirmed significant treatment effect too. Following R code was used to calculate the p-value after the two-stage trial:

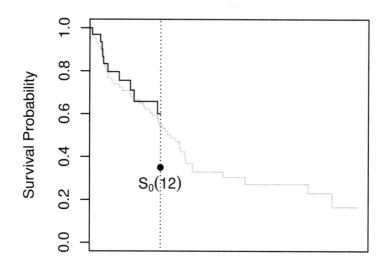

FIGURE 4.2: Survival data for two-stage trial with a restricted follow-up time $x = 12$ months (black).

```
library(mvtnorm)
corr=matrix(c(1,0.5086,0.5086,1),2,2)
pmvnorm(lower=c(0.3225,2.1572),upper=Inf, corr=corr)
```

where the correlation matrix corr is 2×2 matrix with variance 1 and correlation $\rho = 0.5283$.

TABLE 4.4: Restricted follow-up (with restricted follow-up $x = 12$ months) two-stage virtual data ($n = 86$) used in the illustrative data analysis. Entry: time of entry to the trial (months); Time: time to death or to the analysis (months); Δ: censoring indicator (0=censored, 1=died).

Entry	Time	Δ	Entry	Time	Δ	Entry	Time	Δ	Entry	Time	Δ
1.06	12.00	0	10.24	12.00	0	21.28	12.00	0	33.16	1.14	1
1.76	12.00	0	10.78	9.59	1	21.35	12.00	0	33.46	12.00	0
2.10	12.00	0	10.89	12.00	0	21.61	12.00	0	33.75	12.00	0
2.13	3.98	1	12.77	3.28	1	22.74	2.66	1	34.21	9.32	1
2.29	3.28	1	12.86	2.71	1	23.35	2.30	1	34.29	12.00	0
2.55	4.97	1	13.11	12.00	0	23.46	12.00	0	34.99	12.00	0
2.70	8.44	1	13.89	8.13	1	23.46	10.84	1	35.29	12.00	0
3.76	12.00	0	13.91	10.84	1	23.47	12.00	0	36.23	12.00	0
4.36	2.77	1	15.75	12.00	0	23.91	12.00	0	36.52	12.00	0
5.01	7.16	1	15.79	12.00	0	24.40	0.71	1	36.66	2.24	1
5.01	7.33	1	16.04	12.00	0	27.46	12.00	0	36.88	1.86	1
5.67	7.22	1	16.29	10.24	1	27.80	7.93	1	36.88	12.00	0
5.87	1.62	1	17.08	0.73	1	28.77	12.00	0	38.03	12.00	0
6.59	10.96	1	17.39	2.47	1	29.06	0.55	1	39.75	0.86	1
7.01	12.00	0	17.44	10.25	1	29.34	9.58	1	40.60	12.00	0
7.51	4.28	1	17.87	12.00	0	29.97	4.83	1	40.95	1.12	1
7.99	1.61	1	17.95	12.00	0	30.08	0.11	1	41.62	12.00	0
8.32	0.44	1	18.89	12.00	0	30.45	2.40	1	42.80	0.22	1
8.70	7.96	1	19.73	6.67	1	30.92	12.00	0	42.81	1.48	1
9.25	12.00	0	19.75	6.09	1	31.17	0.90	1	42.94	12.00	0
10.07	12.00	0	20.10	7.37	1	31.22	12.00	0			
10.08	12.00	0	21.06	12.00	0	32.05	2.18	1			

TABLE 4.5: Restricted follow-up (with restricted follow-up $x = 12$ months) stage I virtual data ($n_1 = 43$) (with interim time $t_1 = 21.1708$ months) used in the illustrative analysis. Entry: time of entry to the trial (months); Time: time to death or to the analysis (months); Δ: censoring indicator (0=censored, 1=died).

Entry	Time	Δ	Entry	Time	Δ	Entry	Time	Δ	Entry	Time	Δ
1.06	12.00	0	5.67	7.22	1	10.24	10.93	0	16.29	4.88	0
1.76	12.00	0	5.87	1.62	1	10.78	9.59	1	17.08	0.73	1
2.10	12.00	0	6.59	10.96	1	10.89	10.28	0	17.39	2.47	1
2.13	3.98	1	7.01	12.00	0	12.77	3.28	1	17.44	3.73	0
2.29	3.28	1	7.51	4.28	1	12.86	2.71	1	17.87	3.30	0
2.55	4.97	1	7.99	1.61	1	13.11	8.06	0	17.95	3.22	0
2.70	8.44	1	8.32	0.44	1	13.89	7.28	0	18.89	2.28	0
3.76	12.00	0	8.70	7.96	1	13.91	7.26	0	19.73	1.44	0
4.36	2.77	1	9.25	11.92	0	15.75	5.42	0	19.75	1.42	0
5.01	7.16	1	10.07	11.10	0	15.79	5.38	0	20.10	1.07	0
5.01	7.33	1	10.08	11.09	0	16.04	5.13	0			

4.2.7 R Code for Two-Stage Design

Following optimal two-stage design R code 'TwoStage.CM' is modified based on the R package 'Optim2Design' developed by Huang and Thomas (2014). The 'TwoStage.CM' code included four commonly used parametric survival distributions for conducting single-arm optimal two-stage design testing the survival probability at a fixed time point. The four parametric survival distributions implemented in the 'TwoStage.CM' code are the Weibull distribution (WB), log-normal (LN), gamma (GM) and log-logistic (LG) distributions.

In the two-stage design, we assumed that the shape parameters of four distributions are known or can be estimated from historical data and the scale parameters of the corresponding distributions are calculated from $S_0(x)$ and $S_1(x)$ for the null and alternative survival distributions, respectively. For example, for the Weibull distribution $S(x) = e^{-(x/b)^a}$, where a and b are the shape and scale parameters and the scale parameters are solved to be $b_0 = x/\{-\log S_0(x)\}^{1/a}$ and $b_1 = x/\{-\log S_1(x)\}^{1/a}$ for the null and alternative survival distributions, respectively. Therefore, in general the distribution under the alternative does not satisfy the proportional hazard assumption to the null distribution (except for the exponential distribution).

```
######### Two-Stage Design for Evaluating Survival Probability at x ##########
### shape is the shape parameter of the underlying survival distribution;   ###
### S0 and S1 are survival probability at fixed time point x; dist specifies###
### the underlying survival distribution; r is accrual rate per unit time;  ###
### alpha and beta are the type I and type II errors and power=1-beta.       ###
#############################################################################
library(survival)
library(mvtnorm)
TwoStage.CM=function(shape, S0, S1, x, r, alpha, beta, dist)
{
    n0 <- FixDes(r,alpha, beta, shape, S0, S1, x)
    if (x>n0/r) {stop("lanmark time point x is large than study duration!")}
    n.int=1
    nsearch=seq(n0, n0+20, n.int)
    nvec <- length(nsearch)
    nrho.res <- matrix(rep(NA, nvec * 2), nrow = nvec, ncol = 2)
    nrho.res[, 1] <- nsearch
    n.res <- nsearch
    nvec <- length(n.res)
    min.res <- rep(NA, nvec)
    t1.res <-rep(NA, nvec)
    t2.res <-rep(NA, nvec)
    C1.res <-rep(NA, nvec)
    C2.res <-rep(NA, nvec)
    MTSL.res <-rep(NA, nvec)
    MDA.res <-rep(NA, nvec)
    ES.res <-rep(NA, nvec)
    ETSL.res <-rep(NA, nvec)
    PS0.res <-rep(NA, nvec)
    PS1.res <-rep(NA, nvec)
    n1.res <-rep(NA, nvec)
    n2.res <-rep(NA, nvec)
    u.res<-matrix(rep(NA, nvec * 2), nrow = nvec, ncol = 2)
```

```
   se.res<-matrix(rep(NA, nvec * 4), nrow = nvec, ncol = 4)

  tol=10^(-6)

i=0
for (n in nsearch){
  i=i+1
  optout=optimize(f=f.DesWrap, c(0,1), tol=tol, r, alpha, beta,
                  shape, S0, S1, x, n, dist)
          min.res[i] <- optout$objective
          nrho.res[i, 2] <- optout$minimum
          cat("n=", nrho.res[i, 1], "optimal rho=", nrho.res[i,2],
              "ES=", min.res[i], "\n")
          flush.console()

  rho=nrho.res[i,2]
  f.out<-f.Des(rho, r, alpha, beta, shape, S0, S1, x, n, dist)

  MDA.res[i]  <- f.out$mda
  MTSL.res[i] <- f.out$mda+x
  se.res[i,]  <- f.out$se
  u.res[i,]   <- f.out$u
  ETSL.res[i] <- f.out$ETSL
  ES.res[i]   <- f.out$ES
  PS0.res[i]<-f.out$PS0
  PS1.res[i]<-f.out$PS1
  n1.res[i]   <- ceiling(f.out$n1)
  n2.res[i]   <- ceiling(f.out$n2)
  t1.res[i]   <- f.out$t1
  t2.res[i]   <- f.out$t2
  C1.res[i]   <- f.out$C1
  C2.res[i]   <- f.out$C2
}
## truncate result vectors at last sample size evaluated
nrho.res <- nrho.res[1:i,]
min.res <-  min.res[1:i]
n.res <-    n.res[1:i]
t1.res <-   t1.res[1:i]
t2.res <-   t2.res[1:i]
C1.res <-   C1.res[1:i]
C2.res <-   C2.res[1:i]
MTSL.res <- MTSL.res[1:i]
MDA.res <-  MDA.res[1:i]
ETSL.res <- ETSL.res[1:i]
ES.res <-   ES.res[1:i]
PS0.res <-  PS0.res[1:i]
PS1.res <-  PS1.res[1:i]
n1.res <-   n1.res[1:i]
n2.res <-   n2.res[1:i]
se.res <-   se.res[1:i, ]
u.res <-    u.res[1:i, ]

## select the optimal n
order.min <- order(min.res)[1]
n.last <- nrho.res[order.min,1]
outcome <- min.res[order.min]
ETSL<-ETSL.res[order.min]
```

```
  ES <- ES.res[order.min]
  PS0 <- PS0.res[order.min]
  PS1 <- PS1.res[order.min]
  mda <- MDA.res[order.min]
  t1.last <- t1.res[order.min]
  t2.last <- t2.res[order.min]
  C1.last <- C1.res[order.min]
  C2.last <- C2.res[order.min]
  n1 <- n1.res[order.min]
  n2 <- n2.res[order.min]
  se <- se.res[order.min,]
  u  <- u.res[order.min,]

  res=structure(
      list(test=c(alpha=alpha,beta=beta,shape=shape,S0=S0,S1=S1,x=x,n0=n0),
      result=round(c(ES=ES,ETSL=ETSL),4),
              StopProb=round(c(StopProb.Null=PS0,StopProb.Alt=PS1),4),
      n=c(Stage1.n1=n1,FinalMax.n=n.last),
      boundary=round(c(C1=C1.last,C2=C2.last),4),
      stageTime=round(c(t1=t1.last,t2=t2.last,MTSL=mda+x,MDA=mda),4)))
  return(res)
}

FixDes=function(r,alpha,beta,shape,S0, S1, x){
   sig21 <- sqrt((1 - S1)/S1)
       #require all patients be followed x length time#
   lam0 <- -log(S0)
   lam1 <- -log(S1)
   n0 <- (sig21*(qnorm(1-alpha)+qnorm(1-beta)))/
         ((log(lam0)-log(lam1))*lam1))^2
   return(ceiling(n0))
}

f.DesWrap=function(rho, r, alpha, beta, shape, S0, S1, x, n, dist)
{
  out<-f.Des(rho, r, alpha, beta, shape, S0, S1, x, n, dist)
  return(out$min)
}

f.Des=function(rho, r, alpha, beta, shape, S0, S1, x, n, dist)
{
    mda=n/r                  #2

    if (dist=="WB"){
     s=function(a,b,u){1-pweibull(u,a,b)}
     f=function(a,b,u){dweibull(u,a,b)}
     h=function(a,b,u){f(a,b,u)/s(a,b,u)}
     scale0=x/(-log(S0))^(1/shape)
     scale1=x/(-log(S1))^(1/shape)
     }

    if (dist=="LN"){
     s=function(a,b,u){1-plnorm(u,b,a)}
     f=function(a,b,u){dlnorm(u,b,a)}
     h=function(a,b,u){f(a,b,u)/s(a,b,u)}
     scale0=log(x)-shape*qnorm(1-S0)
     scale1=log(x)-shape*qnorm(1-S1)
```

```
}

if (dist=="LG"){
 s=function(a,b,u){1/(1+(u/b)^a)}
 f=function(a,b,u){(a/b)*(u/b)^(a-1)/(1+(u/b)^a)^2}
 h=function(a,b,u){f(a,b,u)/s(a,b,u)}
 scale0=x/(1/S0-1)^(1/shape)
 scale1=x/(1/S1-1)^(1/shape)
}

if (dist=="GM"){
 s=function(a,b,u){1-pgamma(u,a,b)} ## shape=a
 f=function(a,b,u){dgamma(u,a,b)}
 h=function(a,b,u){f(a,b,u)/s(a,b,u)}
 root0=function(t){s(shape,t,x)-S0}
 scale0=uniroot(root0,c(0,10))$root
 root1=function(t){s(shape,t,x)-S1}
 scale1=uniroot(root1,c(0,10))$root
}

sig21 <- sqrt((1 - S1)/S1)            #3
sig20 <- sqrt((1 - S0)/S0)            #3
lam0 <- -log(S0)
lam1 <- -log(S1)
sig11=sig21/rho                       #4
f.int=function(u,r,shape,scale,x,n,t,mda){
  h(shape,scale,u)/(s(shape,scale,u)*(1-punif(u, t-mda, t)))
}
fG=function(t,r,shape,scale,x,n,mda,sig11){
   integrate(f.int,0,x,r,shape,scale,x,n,t,mda)$value-sig11^2
}
epsilon=10^(-5)
t1.root=uniroot(fG,c(x+epsilon,mda+x),r,shape,scale1,x,n,mda,sig11)#5
t1=t1.root$root
t2=max(mda-t1,0)                      #6
g.int=function(u,r,shape,scale,x,n,mda,t1){
  h(shape, scale, u)/(s(shape,scale,u)*(1-punif(u, t1-mda, t1)))
}
sig10=sqrt(integrate(g.int,0,x,r,shape,scale0,x,n,mda,t1)$value) #7
rho0=sig20/sig10                      #8
u2=sqrt(n)*(log(lam0)-log(lam1))*lam1/sig21 #9
u1=rho*u2
u=c(u1,u2)

sigma1=matrix(c(1,rho,rho,1),2,2)
C1low<- -3.5
C1up<-qnorm(1-alpha)
C2.init<- qnorm(1-alpha)
powf=function(C1, C2.init, sigma1,beta,u1,u2)
  {pmvnorm(lower=c((C1-u1),(C2.init-u2)),upper=c(Inf,Inf),
          sigma=sigma1)-(1-beta)}
C1root <- uniroot(powf,c(C1low,C1up),
   C2=C2.init,sigma1=sigma1,beta=beta,u1=u1,u2=u2)
C1.init <- C1root$root
par.init=c(C1.init,C2.init)
f=function(x){
 x1=x[1]
```

```
      x2=x[2]
      A=pmvnorm(lower=c(x1,x2),upper=c(Inf, Inf), mean=c(0,0),
               corr=matrix(c(1,rho0,rho0,1),2,2))-alpha
      B=pmvnorm(lower=c(x1,x2),upper=c(Inf, Inf), mean=c(u1,u2),
               corr=matrix(c(1,rho,rho,1),2,2))-(1-beta)
      A^2+B^2
      }
      fit=optim(par=par.init,f)
      C1=fit$par[1]
      C2=fit$par[2]
      PS0=pnorm(C1)
      PS1=pnorm(C1-u1)
      n1=t1*r
      n2=t2*r
      ETSL=t1+(1-PS0)*(mda+x-t1)
      ES=n1+(1-PS0)*n2        #11

      return(list(min=ES,ETSL=ETSL, ES=ES, PS0=PS0, PS1=PS1, n1=n1, n2=n2,
        t1=t1,t2=t2,mda=mda,C1=C1,C2=C2,se=c(sig10,sig20,sig11,sig21),u=u))
}
TwoStage.CM(shape=1,S0=0.5,S1=0.63,x=1,r=20,alpha=0.05,beta=0.2,dist="WB")
$test
 alpha   beta  shape    S0     S1      x      n0
  0.05   0.20   1.00   0.50   0.63   1.00  104.00


$result
      ES    ETSL
 75.4141  4.1119

$StopProb
StopProb.Null  StopProb.Alt
      0.6588        0.1289

$n
 Stage1.n1 FinalMax.n
        52        122

$boundary
     C1     C2
 0.4092 1.5321

$stageTime
     t1     t2   MTSL    MDA
 2.5644 3.5356 7.1000 6.1000
```

4.3 Two-Stage Design Based on Arcsin-Square Root Test

4.3.1 Test Statistics

Two-stage design based on the arcsin-square transformed root test is discussed in this section. Similar to the log-log transformed test, the following counting process

$$\sqrt{n}\{\arcsin\sqrt{\hat{S}(x;t)} - \arcsin\sqrt{S_0(x)}\}; \quad t > x$$

converges weakly to a Gaussian process $W(x;t)$ with independent increments and variance function

$$\nu^2(x;t) = \frac{S_0(x)\sigma^2(x;t)}{4\{1 - S_0(x)\}},$$

and

$$\sigma^2(x;t) = \int_0^x \frac{\lambda(u)}{\pi(u;t)}du,$$

where $\pi(u;t) = P(X_1(t) \geq u)$, and variance $\sigma^2(x;t)$ can be estimated by

$$\hat{\sigma}^2(x;t) = \int_0^x \frac{\bar{N}(du,t)}{\bar{Y}^2(u,t)/n},$$

and variance $\nu^2(x;t)$ can be estimated by

$$\hat{\nu}^2(x;t) = \frac{\hat{S}(x;t)\hat{\sigma}^2(x;t)}{4\{1 - \hat{S}(x;t)\}}.$$

The arcsin-square root transformed sequential test is defined by

$$Z(x;t) = \frac{\sqrt{n}\{\arcsin\sqrt{\hat{S}(x;t)} - \arcsin\sqrt{S_0(x)}\}}{\hat{\nu}(x;t)}.$$

4.3.2 Optimal Two-Stage Design

To study the asymptotic distribution of the arcsin-square root transformed sequential test, let $g(\cdot)$ be a known function whose derivative $g'(\cdot)$ is nonzero and continuous. By functional delta-method, the transformed process

$$\sqrt{n}\left[g\{\Lambda_0(x)\} - g\{\hat{\Lambda}(x;t)\}\right], \quad t > x$$

converges weakly to a zero-mean Gaussian process $\bar{W}(x;t)$ with

$$\mathrm{cov}\{\bar{W}(x;t), \bar{W}(y;s)\} \simeq g'\{\Lambda(x)\}g'\{\Lambda(y)\}\mathrm{cov}\{W(x;t), W(y;s)\},$$

where $W(x;t)$ is the zero-mean Gaussian process of $\sqrt{n}\{\Lambda_0(x) - \hat{\Lambda}(x;t)\}$. Thus, corr$\{\bar{W}(x;t), \bar{W}(y;s)\} = $ corr$\{W(x;t), W(y;s)\}$. That is the correlation of the asymptotic joint distribution of $\bar{W}(x;t)$ will be the same regardless of the choice of $g(\cdot)$. Therefore, the joint distribution of $Z(x;t)$ and $Z(x;s)$ is asymptotically bivariate normal with correlation given by

$$\rho_k(t,s) = \frac{\sigma_k(x;s)}{\sigma_k(x;t)}$$

where $t \leq s$ and $k = 0,1$ for under the null $(k = 0)$ and alternative distributions $(k = 1)$. To calculate the asymptotic variance $\sigma_1^2(x;t)$ under the null and alternative hypotheses, assuming subjects were recruited with a uniform distribution over the maximum duration of accrual period MDA and followed for an additional period of length x, and the maximum total study duration MTSL $=$ MDA $+ x$. We further assume no loss to follow-up. Then, under the null hypothesis, we have $\pi(u;t) = P(T_1 \wedge x \wedge (t - A_1)^+ \geq u) = P(T_1 \geq u)P(A_1 < t - u) = S_0(u)P(A_1 < t - u), 0 < u \leq x$. Thus, under the null hypothesis, we have

$$\sigma_0^2(x;t) = \int_0^x \frac{\lambda_0(u)}{S_0(u)P(A_1 < t - u)} du, \quad x < t \leq \text{MTSL}$$

where $\lambda_0(\cdot)$ and $S_0(\cdot)$ are the hazard and survival distribution function at the null, respectively; under the alternative hypothesis, we have

$$\sigma_1^2(x;t) = \int_0^x \frac{\lambda_1(u)}{S_1(u)P(A_1 < t - u)} du, \quad x < t \leq \text{MTSL}$$

where $\lambda_1(\cdot)$ and $S_1(\cdot)$ are the hazard and survival distribution function at the alternative, respectively.

Under the null hypothesis H_0, (Z_1, Z_2) is approximately bivariate normal distributed with mean zero, variance one and correlation $\rho_0 = \sigma_{20}/\sigma_{10}$, where

$$\sigma_{10}^2 = \int_0^x \frac{\lambda_0(u)du}{S_0(u)P(A_1 \leq t_1 - u)}$$

and

$$\sigma_{20}^2 = \int_0^x \frac{\lambda_0(u)du}{S_0(u)P(A_1 \leq \text{MTSL} - u)} = \frac{1}{S_0(x)} - 1.$$

Under alternatives H_1, (Z_1, Z_2) is approximately bivariate normal distribution with mean $\mu = (\rho_1 u, u)$, variance one and correlation $\rho_1 = \sigma_{21}/\sigma_{11}$, where

$$u = \frac{\sqrt{n}\{\arcsin\sqrt{S_1(x)} - \arcsin\sqrt{S_0(x)}\}}{\nu_{21}},$$

with

$$\nu_{21}^2 = \frac{S_1(x)\sigma_{21}^2}{4\{1 - S_1(x)\}},$$

$$\sigma_{11}^2 = \int_0^x \frac{\lambda_1(u)du}{S_1(u)P(A_1 \le t_1 - u)},$$

and

$$\sigma_{21}^2 = \int_0^x \frac{\lambda_1(u)du}{S_1(u)P(A_1 \le \text{MTSL} - u)} = \frac{1}{S_1(x)} - 1.$$

4.3.3   Study Design and R Code

To illustrate two-stage trial design using the arcsin-square root test, as in the study design given in the previous section for testing the hypothesis that the 12-month survival probability of $S_0(12) = 35\%$ or less on treatment is considered inefficient, while 12-month survival probability of $S_1(12) = 50\%$ or greater would be worthwhile, we assume: uniform accrual with accrual rate 2 subjects per month and 12 months follow-up; survival times of experimental group follow an exponential distribution $S(t) = e^{-\lambda t}$ with parameter $\lambda = -\log S(12)/12 = -\log(0.5)/12$. With type I error rate $\alpha = 0.05$ and power of 80%, an optimal two-stage design can be derived from R code 'TwoStage.AS' which refers to optimal two-stage based on arcsin-square root test.

Using R code 'TwoStage.AS', we obtain the following results:

```
TwoStage.AS(shape=1,S0=0.35,S1=0.5,x=12,r=2,alpha=0.05,beta=0.2,dist="WB")
$test
alpha  beta shape   S0    S1    x    n0
 0.05  0.20  1.00  0.35  0.50 12.00 67.00

$result
      ES      ETSL
53.42247 31.33645

$StopProb
StopProb.Null  StopProb.Alt
   0.6145652     0.1142643

$n
  Stage1.n1 FinalMax.n
    40            76

$boundary
        C1         C2
0.2912376 1.5512701

$stageTime
      t1       t2    MTSL     MDA
19.6313 18.3687 50.0000 38.0000
```

This two-stage design can be executed as follows. At stage 1, we will enroll 40 patients and the interim analysis will be performed after the 40th patient is enrolled in the study or at calendar time $t_1 = 19.6313$ (months), whichever occurs later. The stage 1 test statistic $Z_1(x; t_1)$ will be calculated and compared with stage 1 boundary $c_1 = 0.2912$. If $Z_1(x; t_1) < c_1$, the trial will stop for futility, otherwise, the trial will go to second stage. At stage 2, an additional 36 patients will be enrolled in the study for a total of 76 patients. Once the final patient is enrolled and followed for $x = 1$ year, the final analysis will be conducted to calculate the final test statistic $Z_2(x; \text{MTSL})$. If $Z_2(x; \text{MTSL}) < c_2 = 1.5513$, we will conclude that the new treatment is not worth further investigation. If $Z_2(x; \text{MTSL}) > c_2$, we will conclude that the new treatment is promising. The operating characteristics of this two-stage design are summarized in Table 6.3. The expected sample size for the two-stage design is 54 and expected total study length is 31.34 months; the first stage stopping probabilities at null and alternative are 0.6146 and 0.1143, respectively.

The R code 'TwoStage.AS' implemented optimal two-stage design using arcsin-square root test is given in Appendix K.

4.3.4 Simulation

To study operating characteristics of the proposed optimal two-stage design based on arcsin-square root transformed test, we conducted simulation under various scenarios. In the simulations, the survival distribution was taken as the Weibull distribution with shape parameter 0.5, 1 and 2; the survival probability under null $S_0(1)$ is set to be 0.3 to 0.6 by 0.1; the scale parameters under the null are determined by the value of $S_0(1)$; the survival probability under alternative $S_1(1)$ is set to be 0.5 to 0.8 by 0.1, or a 20% increase in survival probability at $x = 1$; accrual rate is set to be $r = 10$; type I error rate $\alpha = 0.05$ and power of 80%. Under each scenario, the R code 'TwoStage.AS' was used to create the two-stage designs. Based on the output parameters of each two-stage design, we conducted a simulation study to simulate the empirical type I error and empirical power. The simulation results for the Nelson-Aalen estimate based test and Kaplan-Meier estimate based test are presented in Table 4.6. From simulations we have the following observations: The two-stage design for both Nelson-Aalen based tests did not preserve the type I error well but provided adequate power. The Kaplan-Meier based test preserved both type I error and power. The two-stage designs were quite robust across the different shape parameters of the Weibull distribution.

TABLE 4.6: Optimal two-stage characteristics for designs based on arcsin-square root transformed test under Weibull distribution. Overall empirical type I error rate ($\hat{\alpha}$) and empirical power (EP) for the two-stage designs were estimated from 10,000 simulated trials.

Shape	(S_0, S_1)	t_1	c_1	c_2	ES	ETSL	n_1	n_{max}	PS_0	PS_1	$\hat{\alpha}$	EP
				Nelson-Aalen Estimate Based Tests								
0.5	(.3, .5)	1.87	.414	1.528	26.9	3.03	19	43	.66	.13	.055	.83
	(.4, .6)	1.92	.422	1.528	27.9	3.13	20	45	.66	.13	.072	.84
	(.5, .7)	1.85	.420	1.530	26.8	3.02	19	43	.66	.13	.053	.83
	(.6, .8)	1.66	.408	1.534	23.6	2.70	17	37	.66	.13	.063	.83
1	(.3, .5)	1.97	.375	1.532	27.9	3.15	20	43	.65	.13	.055	.83
	(.4, .6)	2.02	.384	1.531	28.9	3.24	21	45	.65	.13	.073	.84
	(.5, .7)	1.96	.383	1.533	27.8	3.13	20	43	.65	.13	.052	.83
	(.6, .8)	1.76	.362	1.537	24.6	2.81	18	37	.64	.12	.063	.83
2	(.3, .5)	2.06	.301	1.548	28.8	3.26	21	42	.62	.12	.050	.81
	(.4, .6)	2.12	.314	1.547	29.8	3.35	22	44	.62	.12	.060	.83
	(.5, .7)	2.05	.307	1.549	28.7	3.24	21	42	.62	.12	.068	.84
	(.6, .8)	1.85	.267	1.556	25.4	2.94	19	36	.61	.11	.072	.85
				Kaplan-Meier Estimate Based Tests								
0.5	(.3, .5)	1.87	.414	1.528	26.9	3.03	19	43	.66	.13	.052	.81
	(.4, .6)	1.92	.422	1.528	27.9	3.13	20	45	.66	.13	.043	.77
	(.5, .7)	1.85	.420	1.530	26.8	3.02	19	43	.66	.13	.051	.80
	(.6, .8)	1.66	.408	1.534	23.6	2.70	17	37	.66	.13	.060	.81
1	(.3, .5)	1.97	.375	1.532	27.9	3.15	20	43	.65	.13	.052	.80
	(.4, .6)	2.02	.384	1.531	28.9	3.24	21	45	.65	.13	.043	.78
	(.5, .7)	1.96	.383	1.533	27.8	3.13	20	43	.65	.13	.050	.81
	(.6, .8)	1.76	.362	1.537	24.6	2.81	18	37	.64	.12	.060	.81
2	(.3, .5)	2.06	.301	1.548	28.8	3.26	21	42	.62	.12	.049	.79
	(.4, .6)	2.12	.314	1.547	29.8	3.35	22	44	.62	.12	.057	.81
	(.5, .7)	2.05	.307	1.549	28.7	3.24	21	42	.62	.12	.065	.82
	(.6, .8)	1.85	.267	1.556	25.4	2.94	19	36	.61	.11	.037	.76

5

One-Stage Design Evaluating Survival Distributions

5.1 Introduction

There are several major disadvantages for the methods discussed in Chapters 3 and 4. First, the nonparametric cumulative hazard based test statistics do not preserve the type I error and power well, particularly when sample size is small, which is typically the case in phase II trials. Second, it is difficult to chose the landmark time point. Third, events that occurred after the landmark time point are not used in the test, which results in an inefficient study design. A more efficient study design is to compare the survival distribution of the experimental group to the survival distribution of the reference group, which is also referred to as the null survival distribution. The null survival distribution can be estimated from historical data of the standard of care; however, it is treated as fixed and is not subject to the variation for the purpose of single-arm phase II trial design. Thus, the one-sample log-rank test (OSLRT) can be used for the trial design and data analysis.

The hypothesis testing for such trial design can be formulated as follows. Let $\Lambda_0(x)$ and $S_0(x)$ be the known cumulative hazard function and survival function, respectively, under the null hypothesis, and let $\Lambda(x)$ and $S(x)$ be the unknown cumulative hazard and survival functions for the new treatment. The study may consider testing the following hypothesis of interest (Figure 5.1) using the OSLRT:

$$H_0 : S(x) \le S_0(x) \quad \text{vs} \quad H_1 : S(x) > S_0(x), \quad \text{for all} \ x > 0.$$

The OSLRT is the difference between the observed number of events and the expected number of events under the null survival distribution. It was first introduced by Breslow (1975). The application of the OSLRT to the single-arm phase II trial designs was discussed by Finkelstein et al. (2003), Sun et al. (2011), Kwak and Jung (2013), Wu (2014, 2015), Schmidt et al. (2015), Belin et al. (2017), Wu et al. (2019b) and Chu et al. (2020). In this chapter, the OSLRT is introduced and sample size formula is derived for single-arm phase II trial design with time-to-event endpoint.

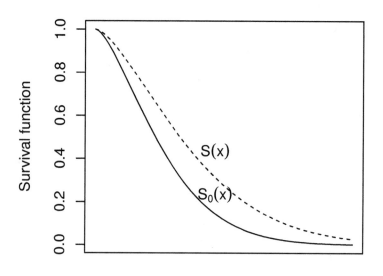

FIGURE 5.1: Hypothesis testing for evaluating survival distributions.

5.2 One-Sample Log-Rank Test

5.2.1 Test Statistics

To introduce the OSLRT, we assume that the failure time T_i and censoring time C_i are independent and $\{T_i, C_i, i = 1, \ldots, n\}$ are independent and identically distributed. The observed failure time and failure indicator are $X_i = T_i \wedge C_i$ and $\Delta_i = I(T_i \leq C_i)$, respectively, for the ith subject. On the basis of the observed data $\{X_i, \Delta_i, i = 1, \cdots, n\}$, we define $O = \sum_{i=1}^{n} \Delta_i$ as the observed number of events, and $E = \sum_{i=1}^{n} \Lambda_0(X_i)$ as the expected number of events, where $\Lambda_0(x)$ is the cumulative hazard function under the null hypothesis. Then, the OSLRT is defined by

$$L_1 = \frac{E - O}{\sqrt{E}}. \tag{5.1}$$

To study the asymptotic distribution of the OSLRT, we formulate it using counting process notations. Specifically, let $N_i(x) = \Delta_i I\{X_i \leq x\}$ and $Y_i(x) = I\{X_i \geq x\}$ be the failure and at-risk processes, respectively, then

$$O = \sum_{i=1}^{n} \int_0^{\infty} dN_i(x), \quad E = \sum_{i=1}^{n} \int_0^{\infty} Y_i(x) d\Lambda_0(x).$$

Thus, the counting process formulation of the OSLRT is given by

$$L_1 = \frac{U}{\hat{\sigma}},$$

where

$$U = n^{-1/2} \sum_{i=1}^{n} \int_0^{\infty} \{Y_i(x) d\Lambda_0(x) - dN_i(x)\},$$

and

$$\hat{\sigma}^2 = n^{-1} \sum_{i=1}^{n} \int_0^{\infty} Y_i(x) d\Lambda_0(x).$$

Under the null hypothesis H_0, by the law of large numbers, $n^{-1} \sum_{i=1}^{n} Y_i(x)$ uniformly converges to $P(X_i \geq x)$, where $P(X_i \geq x) = G(x)S_0(x)$, and $G(x)$ is the survival distribution of censoring time C_i. Thus, $\hat{\sigma}^2$ converges in probability to $\int_0^{\infty} G(x)S_0(x)d\Lambda_0(x)$, which is the exact variance of U under the null (see Appendix C). As U is the sum of independent random variables, thus, by the central limit theorem, U is asymptotically normal distributed under the null hypothesis. Finally, by Slutsky's theorem, $L_1 = U/\hat{\sigma}$ is asymptotically standard normal distributed under the null hypothesis. Hence, we reject null hypothesis H_0 with one-sided type I error rate α if $L_1 = U/\hat{\sigma} > z_{1-\alpha}$, where $z_{1-\alpha}$ is the $100(1-\alpha)$ percentile of the standard normal distribution.

Remark 1: We reversed the order O and E in the test statistic L_1 so that reducing the number of observed events indicates a positive value of L_1.

5.2.2 Sample Size Formula

To design the study, sample size must be calculated to detect a pre-specified alternative survival distribution $S_1(x)$, given the type I error rate α and power $1 - \beta$. Finkelstein et al. (2003) provided a sample size formula for the OSLRT. Unfortunately, the formula was recorded in error. Kwak and Jung (2013) derived a sample size formula for the OSLRT using a naive variance estimate under the alternative which results in an underestimated sample size. To provide accurate sample size estimation, Wu (2015) derived an exact variance for the OSLRT and its asymptotic distribution under the alternative.

Under alternative H_1, $n^{-1}\sum_{i=1}^{n} Y_i(x)$ uniformly converges to $G(x)S_1(x)$, so that

$$\hat{\sigma}^2 \to \sigma_0^2 = \int_0^\infty G(x)S_1(x)d\Lambda_0(x),$$

where $S_1(x) = \exp\{-\Lambda_1(x)\}$ is the survival distribution under the alternative. On the other hand, under the alternative, U is approximately normal with mean $\sqrt{n}\omega = \sqrt{n}(\sigma_0^2 - \sigma_1^2)$, where $\sigma_1^2 = \int_0^\infty G(x)S_1(x)d\Lambda_1(x)$ (see Appendix C for the derivation).

To derive the sample size formula, let the exact mean and variance of W at alternative hypothesis be $E_{H_1}(U) = \sqrt{n}\omega$ and $\text{Var}_{H_1}(U) = \sigma^2$, respectively, then the power of the OSLRT $L_1 = U/\hat{\sigma}$ should satisfy the following equations:

$$
\begin{aligned}
1 - \beta = P(L_1 > z_{1-\alpha}) &\simeq P\left(\frac{U}{\sigma_0} > z_{1-\alpha}|H_1\right) \\
&\simeq P\left(\frac{U - \sqrt{n}\omega}{\sigma} > \frac{\sigma_0}{\sigma}z_{1-\alpha} - \frac{\sqrt{n}\omega}{\sigma}|H_1\right) \\
&\simeq \Phi\left(-\frac{\sigma_0}{\sigma}z_{1-\alpha} + \frac{\sqrt{n}\omega}{\sigma}\right).
\end{aligned}
$$

Therefore, the required sample size is given by

$$n = \frac{(\sigma_0 z_{1-\alpha} + \sigma z_{1-\beta})^2}{\omega^2}, \tag{5.2}$$

where

$$
\begin{aligned}
\omega &= \sigma_0^2 - \sigma_1^2 & (5.3) \\
\sigma^2 &= v_1 - v_1^2 + 2v_{00} - v_0^2 - 2v_{01} + 2v_0 v_1 & (5.4)
\end{aligned}
$$

with v_0, v_1, v_{00} and v_{01} given as follows:

$$v_0 = \sigma_0^2 = \int_0^\infty G(x)S_1(x)d\Lambda_0(x), \tag{5.5}$$

$$v_1 = \sigma_1^2 = \int_0^\infty G(x)S_1(x)d\Lambda_1(x), \tag{5.6}$$

$$v_{00} = \int_0^\infty G(x)S_1(x)\Lambda_0(x)d\Lambda_0(x), \tag{5.7}$$

$$v_{01} = \int_0^\infty G(x)S_1(x)\Lambda_0(x)d\Lambda_1(x), \tag{5.8}$$

The details of derivation are given in Appendix C.

If the underlying model is a proportional hazards model with hazard ratio δ; then the survival distribution and cumulative hazards function under the alternative are determined as $S_1(x) = [S_0(x)]^\delta$ and $\Lambda_1(x) = \delta\Lambda_0(x)$, respectively. Therefore sample size depends only on the hazard ratio δ, survival distribution $S_0(x)$ under the null hypothesis and survival distribution $G(x)$ of censoring time.

Example 4 *Continuation of Trial for Rhabdoid Tumor*

The Weibull distribution was fitted well to the relapse-free survival (RFS) Kaplan-Meier curve of the St. Jude's rhabdoid tumor data with shape parameter $\kappa = 1.32$ and a median RFS time of $m_0 = 0.83$ years, which determines the reference survival distribution $S_0(x) = e^{-\log(2)(\frac{t}{m_0})^\kappa}$ for the single-arm phase II trial design for testing the null hypothesis $H_0 : S(x) = S_0(x)$ for all $x > 0$. The study is powered at alternative $S_1(x) = e^{-\log(2)(\frac{t}{m_1})^\kappa}$ with median survival time $m_1 = 1.2$. The corresponding survival probabilities at $x = 1$ under the null and alternative are $S_0(1) = 0.412$ and $S_1(1) = 0.580$, respectively. Thus, given power of 90%, one-sided type I error rate of 10%, two years accrual and additional one year follow-up, using R function 'Size', the required sample size is $n = 41$. From 10,000 simulations, the estimated empirical type I error and power for the trial are 0.088 and 92%, respectively.

```
########## Sample Size Calculation for OSLRT under Various Distributions ######
### shape is the Weibull shape parameter; S0 and S1 are survival probability###
### at time point x under null and alternative; alpha and beta are type I   ###
### and type II errors; ta and tf are the accrual and follow-up time;       ###
### dist option is to specify one of four parametric survival distributions ###
################################################################################
Size=function(shape,S0,S1,x,ta,tf,dist,alpha,beta)
{tau=ta+tf
 if (dist=="WB"){
  s=function(a,b,u){1-pweibull(u,a,b)}
  f=function(a,b,u){dweibull(u,a,b)}
  h=function(a,b,u){f(a,b,u)/s(a,b,u)}
  H=function(a,b,u){-log(s(a,b,u))}
  scale0=x/(-log(S0))^(1/shape)
  scale1=x/(-log(S1))^(1/shape)}
```

```
if (dist=="LN"){
 s=function(a,b,u){1-plnorm(u,b,a)}
 f=function(a,b,u){dlnorm(u,b,a)}
 h=function(a,b,u){f(a,b,u)/s(a,b,u)}
 H=function(a,b,u){-log(s(a,b,u))}
 scale0=log(x)-shape*qnorm(1-S0)
 scale1=log(x)-shape*qnorm(1-S1)}

if (dist=="LG"){
 s=function(a,b,u){1/(1+(u/b)^a)}
 f=function(a,b,u){(a/b)*(u/b)^(a-1)/(1+(u/b)^a)^2}
 h=function(a,b,u){f(a,b,u)/s(a,b,u)}
 H=function(a,b,u){-log(s(a,b,u))}
 scale0=x/(1/S0-1)^(1/shape)
 scale1=x/(1/S1-1)^(1/shape)}

if (dist=="GM"){
 s=function(a,b,u){1-pgamma(u,a,b)} ## shape=a; scale=b
 f=function(a,b,u){dgamma(u,a,b)}
 h=function(a,b,u){f(a,b,u)/s(a,b,u)}
 H=function(a,b,u){-log(s(a,b,u))}
 root0=function(t){s(shape,t,x)-S0}
 scale0=uniroot(root0,c(0,10))$root
 root1=function(t){s(shape,t,x)-S1}
 scale1=uniroot(root1,c(0,10))$root}

G=function(t){1-punif(t, tf, tau)}
f0=function(t){s(shape,scale1,t)*h(shape,scale0,t)*G(t)}
f1=function(t){s(shape,scale1,t)*h(shape,scale1,t)*G(t)}
f2=function(t){s(shape,scale1,t)*h(shape,scale0,t)*H(shape,scale0,t)*G(t)}
f3=function(t){s(shape,scale1,t)*h(shape,scale1,t)*H(shape,scale0,t)*G(t)}

z0=qnorm(1-alpha)
z1=qnorm(1-beta)
v0=integrate(f0, 0, tau)$value
v1=integrate(f1, 0, tau)$value
v00=integrate(f2, 0, tau)$value
v01=integrate(f3, 0, tau)$value
n=(sqrt(v0)*z0+sqrt(v1+2*v0*v1-v0^2-v1^2+2*v00-2*v01)*z1)^2/(v1-v0)^2
ans=ceiling(n)
return(ans)
}
Size(shape=1.32,S0=0.412,S1=0.580,x=1,ta=2,tf=1,dist="WB",alpha=0.1,beta=0.1)
41

############# Calculate OSLRT Test Statistic and Its p-value  ##############
### shape is the Weibull shape parameter; m0 is the median survival time   ###
### data is input survival data in data frame with time variable and cens  ###
### variable with 1-event 0-censoring.                                     ###
############################################################################
OSLRT=function(shape, m0, data)
{
  scale=m0/(-log(0.5))^(1/shape)
  S=function(shape,scale,u){1-pweibull(u,shape,scale)} # null Weibull dist.
  H=function(shape,scale,u){-log(S(shape,scale,u))} # null cumu hazard function
  X=data$time   # observed failure time
  delta=data$cens  # censoring indicate 1-event 0-censoring
```

```
O=sum(delta)           # observed number of events
M=H(shape, scale, X)
E=sum(M)
Z<-(E-O)/sqrt(E)
pvalue=round(1-pnorm(Z),4)
ans=list(Z=round(Z,4), pvalue=pvalue)
return(ans)
}
time=c(3.241,0.559,0.830,0.156,0.148,0.414,1.181,
       0.827,1.501,1.044,1.556,1.225,0.293)
cens=rep(1,13)
data=data.frame(time=time,cens=cens)
OSLRT(shape=1.32, m0=0.83, data)
$Z
[1] -0.0322
$pvalue
[1] 0.5128
```

5.2.3 Accrual Duration Calculation

Given accrual rate r, accrual duration t_a can be solved from the sample size formula by the following equation using R function 'uniroot'

$$\text{root}(t_a) = t_a r - \frac{\{\sigma_0(t_a)z_{1-\alpha} + \sigma(t_a)z_{1-\beta}\}^2}{\omega(t_a)^2},$$

where $\omega(t_a)$, $\sigma^2(t_a)$ and $\sigma_0^2(t_a)$ are given in equations (5.3), (5.4) and (5.5), respectively, emphasized as the functions of accrual duration t_a. The final sample size is given by $n = [rt_a]^+$.

```
########## Duration calculation for OSLRT under various distributions #########
### shape is the Weibull shape parameter; S0 is survival probability at     ###
### fixed time point $x$ under; hr is the hazard ratio; alpha and beta are   ###
### type I and type II errors; ta and tf are the accrual and follow-up time ###
### dist option is to specify one of four parametric survival distributions ###
###############################################################################
Duration=function(shape,S0,x0,hr,tf,rate,alpha,beta,dist)
 {
  if (dist=="WB"){
    s0=function(u){1-pweibull(u,shape,scale0)}
    f0=function(u){dweibull(u,shape,scale0)}
    h0=function(u){f0(u)/s0(u)}
    H0=function(u){-log(s0(u))}
    s=function(b,u) {s0(u)^b}
    h=function(b,u){b*f0(u)/s0(u)}
    H=function(b,u){-b*log(s0(u))}
    scale0=x0/(-log(S0))^(1/shape)
    scale1=hr
  }

  if (dist=="LN"){
    s0=function(u){1-plnorm(u,scale0,shape)}
    f0=function(u){dlnorm(u,scale0,shape)}
    h0=function(u){f0(u)/s0(u)}
```

```
      H0=function(u){-log(s0(u))}
      s=function(b,u) {s0(u)^b}
      h=function(b,u){b*f0(u)/s0(u)}
      H=function(b,u){-b*log(s0(u))}
      scale0=log(x0)-shape*qnorm(1-S0)
      scale1=hr
    }

  if (dist=="LG"){
      s0=function(u){1/(1+(u/scale0)^shape)}
      f0=function(u){(shape/scale0)*(u/scale0)^(shape-1)/(1+(u/scale0)^shape)^2}
      h0=function(u){f0(u)/s0(u)}
      H0=function(u){-log(s0(u))}
      s=function(b,u) {s0(u)^b}
      h=function(b,u){b*f0(u)/s0(u)}
      H=function(b,u){-b*log(s0(u))}
      scale0=x0/(1/S0-1)^(1/shape)
      scale1=hr
    }

  if (dist=="GM"){
    s0=function(u){1-pgamma(u,shape,scale0)}
    f0=function(u){dgamma(u,shape,scale0)}
    h0=function(u){f0(u)/s0(u)}
    H0=function(u){-log(s0(u))}
    s=function(b,u) {s0(u)^b}
    h=function(b,u){b*f0(u)/s0(u)}
    H=function(b,u){-b*log(s0(u))}
    root0=function(t){1-pgamma(x0,shape,t)-S0}
    scale0=uniroot(root0,c(0,10))$root
    scale1=hr
    }

    root=function(ta){
      tau=ta+tf
      G=function(t){1-punif(t, tf, tau)}
      g0=function(t){s(scale1,t)*h0(t)*G(t)}
      g1=function(t){s(scale1,t)*h(scale1,t)*G(t)}
      g00=function(t){s(scale1,t)*H0(t)*h0(t)*G(t)}
      g01=function(t){s(scale1,t)*H0(t)*h(scale1,t)*G(t)}
      p0=integrate(g0, 0, tau)$value
      p1=integrate(g1, 0, tau)$value
      p00=integrate(g00, 0, tau)$value
      p01=integrate(g01, 0, tau)$value
      s1=sqrt(p1-p1^2+2*p00-p0^2-2*p01+2*p0*p1)
      s0=sqrt(p0)
      om=p0-p1
      rate*ta-(s0*qnorm(1-alpha)+s1*qnorm(1-beta))^2/om^2
      }
      tasingle=uniroot(root, lower=0, upper=5*tf)$root
      n<-ceiling(tasingle*rate)
      ta=round(tasingle,4)
      ans=list(ta=ta, n=n)
      return(ans)
    }
Duration(shape=1.22,S0=0.713,x0=5,hr=1/1.75,tf=3,rate=17.6,
    alpha=0.05,beta=0.2,dist="WB")
```

```
$ta
[1] 4.9741
$n
[1] 88
```

5.2.4 Simulation

We conducted simulations to study the performance of the OSLRT under different scenarios. In simulation studies, the survival distribution of
the reference group was taken as the Weibull, log-normal, gamma and log-
logistic. The survival probabilities at $x = 1$ (year) under null were set to be
$S_0(x) = 0.3, 04, 0.5, 0.6$ and to detect a 20% increase for the alternative hypothesis. We assumed a proportional hazards model $S_1(u) = [S_0(u)]^\delta$, where
δ is determined by $\delta = \log S_1(x)/\log S_0(x)$. We further assumed that subjects
were recruited with a uniform distribution over the accrual period $t_a = 2$
(years) and follow-up period $t_f = 1$ (year); no subject was lost to follow-
up; then, the censoring time is uniformly distributed on interval $[t_f, t_a + t_f]$.
Therefore, given the type I error rate 0.05 and power of 80%, the required
sample sizes for each design scenario were calculated using R code 'Size'. For
each calculated sample size n, 100,000 samples were generated to estimate
empirical type I error and empirical power, which were recorded in Table 5.1.
The simulation results showed that the empirical type I error was always below the nominal level 0.05, which indicated that the OSLRT is a conservative
test. The empirical power of the OSLRT was over the nominal level of 80%.
Therefore sample size calculated from formula (5.2) is slightly overestimated.

```
Size=function(shape,S0,S1,x,ta,tf,dist,alpha,beta)
{tau=ta+tf
 if (dist=="WB"){
  s=function(a,b,u){1-pweibull(u,a,b)}
  f=function(a,b,u){dweibull(u,a,b)}
  h=function(a,b,u){f(a,b,u)/s(a,b,u)}
  H=function(a,b,u){-log(s(a,b,u))}
  scale0=x/(-log(S0))^(1/shape)
  scale1=x/(-log(S1))^(1/shape)}

 if (dist=="LN"){
  s=function(a,b,u){1-plnorm(u,b,a)}
  f=function(a,b,u){dlnorm(u,b,a)}
  h=function(a,b,u){f(a,b,u)/s(a,b,u)}
  H=function(a,b,u){-log(s(a,b,u))}
  scale0=log(x)-shape*qnorm(1-S0)
  scale1=log(x)-shape*qnorm(1-S1)}

 if (dist=="LG"){
  s=function(a,b,u){1/(1+(u/b)^a)}
  f=function(a,b,u){(a/b)*(u/b)^(a-1)/(1+(u/b)^a)^2}
  h=function(a,b,u){f(a,b,u)/s(a,b,u)}
  H=function(a,b,u){-log(s(a,b,u))}
  scale0=x/(1/S0-1)^(1/shape)
```

```
    scale1=x/(1/S1-1)^(1/shape)}

if (dist=="GM"){
 s=function(a,b,u){1-pgamma(u,a,b)} ## shape=a; scale=b
 f=function(a,b,u){dgamma(u,a,b)}
 h=function(a,b,u){f(a,b,u)/s(a,b,u)}
 H=function(a,b,u){-log(s(a,b,u))}
 root0=function(t){s(shape,t,x)-S0}
 scale0=uniroot(root0,c(0,10))$root
 root1=function(t){s(shape,t,x)-S1}
 scale1=uniroot(root1,c(0,10))$root}


G=function(t){1-punif(t, tf, tau)}
f0=function(t){s(shape,scale1,t)*h(shape,scale0,t)*G(t)}
f1=function(t){s(shape,scale1,t)*h(shape,scale1,t)*G(t)}
f2=function(t){s(shape,scale1,t)*h(shape,scale0,t)*H(shape,scale0,t)*G(t)}
f3=function(t){s(shape,scale1,t)*h(shape,scale1,t)*H(shape,scale0,t)*G(t)}

z0=qnorm(1-alpha)
z1=qnorm(1-beta)
v0=integrate(f0, 0, tau)$value
v1=integrate(f1, 0, tau)$value
v00=integrate(f2, 0, tau)$value
v01=integrate(f3, 0, tau)$value
n=(sqrt(v0)*z0+sqrt(v1+2*v0*v1-v0^2-v1^2+2*v00-2*v01)*z1)^2/(v1-v0)^2
ans=ceiling(n)
return(ans)
}
Size(shape=0.5,S0=0.3,S1=0.5,x=1,ta=3,tf=1,dist="WB",alpha=0.05,beta=0.2)
31
Size(shape=0.5,S0=0.3,S1=0.5,x=1,ta=3,tf=1,dist="LN",alpha=0.05,beta=0.2)
32
Size(shape=0.5,S0=0.3,S1=0.5,x=1,ta=3,tf=1,dist="GM",alpha=0.05,beta=0.2)
28
Size(shape=0.5,S0=0.3,S1=0.5,x=1,ta=3,tf=1,dist="LG",alpha=0.05,beta=0.2)
34
```

TABLE 5.1: Sample size (n) was calculated using formula (5.2) under Weibull, log-normal, gamma and log-logistic distribution. The corresponding empirical type I error ($\hat{\alpha}$) and empirical power (EP) were estimated based on 10,000 simulation runs.

Design parameters			Weibull			Log-normal			Gamma			Log-logistic		
Shape	$S_0(1)$	$S_1(1)$	n	$\hat{\alpha}$	EP	n	$\hat{\alpha}$	EP	n	$\hat{\alpha}$	EP	n	$\hat{\alpha}$	EP
0.5	.3	.5	31	.040	.825	32	.039	.829	28	.038	.829	34	.040	.829
	.4	.6	32	.041	.823	35	.040	.834	28	.039	.827	37	.041	.824
	.5	.7	30	.040	.819	33	.040	.834	26	.038	.825	35	.040	.815
	.6	.8	26	.039	.823	27	.039	.836	23	.038	.834	30	.039	.815
1	.3	.5	28	.037	.837	32	.038	.830	28	.038	.836	34	.040	.822
	.4	.6	27	.039	.830	34	.038	.823	27	.039	.833	37	.040	.823
	.5	.7	24	.038	.829	33	.038	.828	24	.038	.828	34	.040	.818
	.6	.8	20	.037	.837	27	.037	.821	20	.037	.839	28	.040	.826
2	.3	.5	26	.037	.840	32	.037	.827	28	.039	.829	36	.040	.825
	.4	.6	24	.036	.838	35	.037	.825	28	.039	.834	39	.040	.828
	.5	.7	20	.035	.843	34	.035	.822	25	.039	.841	35	.039	.821
	.6	.8	15	.033	.856	29	.034	.816	19	.038	.836	27	.039	.827

5.3   Modified One-Sample Log-Rank Test

The simulation results in the previous section showed that the OSLRT L_1 is a conservative test and sample size calculated from the formula (5.2) is overestimated (also see Kwak and Jung, 2013; Wu, 2015). This is partly because of the skewness of its distribution. To correct the conservativeness of the OSLRT and provide accurate sample size estimation, Wu (2014) proposed a modified OSLRT (MOSLRT) which is discussed in the next subsection.

5.3.1   Test Statistics

The skewness of the OSLRT is partly attributed to the correlation between the numerator and the denominator of the OSLRT. To reduce the correlation and correct the skewness, we propose a MOSLRT, which is defined as follows:

$$L_2 = \frac{E - O}{\sqrt{(O + E)/2}}.$$

The counting-process formulation of the MOSLRT is given by

$$L_2 = \frac{U}{\hat{\sigma}},$$

where

$$U = n^{-1/2} \sum_{i=1}^{n} \int_0^\infty \{Y_i(t)d\Lambda_0(t) - dN_i(t)\},$$

and

$$\hat{\sigma}^2 = n^{-1} \sum_{i=1}^{n} \int_0^\infty \{dN_i(t) + Y_i(t)d\Lambda_0(t)\}/2.$$

Under the null hypothesis, we have $n^{-1}O \to E_{H_0}(\Delta) = E_{H_0}(\Lambda_0(X))$ and $n^{-1}E \to E_{H_0}(\Lambda_0(X)) = \int_0^\infty G(t)S_0(t)d\Lambda_0(t)$ and $\hat{\sigma}^2 \to \int_0^\infty G(t)S_0(t)d\Lambda_0(t)$ (Appendix C). Therefore, under the null hypothesis, the MOSLRT L_2 is asymptotically standard normal distributed. Hence, we reject the null hypothesis H_0 with one-sided type I error rate α if $L_2 = U/\hat{\sigma} > z_{1-\alpha}$.

Remark 2: When the new treatment is believed to be no worse than the reference group, it is expected that $O \le E$, or $(O + E)/2 \le E$, thus, $L_2 \ge L_1$ and the MOSLRT is a more powerful test than the OSLRT.

5.3.2   Sample Size Formula

Under the alternative, recall that U is approximately normal with mean $\sqrt{n}\omega = \sqrt{n}(\sigma_0^2 - \sigma_1^2)$, where $\sigma_0^2 = \int_0^\infty G(x)S_1(x)d\Lambda_0(x)$ and $\sigma_1^2 =$

$\int_0^\infty G(x)S_1(x)d\Lambda_1(x)$ (see Appendix C). Further it can show that $\hat{\sigma}^2 \to \bar{\sigma}^2 = (\sigma_0^2 + \sigma_1^2)/2$. Let the exact variance of U at alternative hypothesis be $\text{Var}_{H_1}(U) = \sigma^2$. Then, the power of the MOSLRT L_2 should satisfy the following equations:

$$1 - \beta = P(L_2 > z_{1-\alpha}|H_1) \simeq \Phi\left(-\frac{\bar{\sigma}}{\sigma}z_{1-\alpha} + \frac{\sqrt{n}\omega}{\sigma}\right).$$

Therefore, the required sample size for the test statistic L_2 is given by

$$n = \frac{(\bar{\sigma}z_{1-\alpha} + \sigma z_{1-\beta})^2}{\omega^2} \tag{5.9}$$

where σ^2 is given by equation (5.4).

5.3.3 Comparison

To compare performance of the OSLRT and MOSLRT, we conducted simulations under the exponential model. The design parameters for the simulation were set as follows: accrual time $t_a = 1, 3$; follow-up time $t_f = 1, 2, 3$; type I error rate $\alpha = 0.05$ and power $1 - \beta = 80\%$. The survival probability under the null $S_0(x)$ was set to 0.2 to 0.7, and under the alternative $S_1(x)$ was set to 0.35 to 0.8, where $x = 1, 2$. Under these design scenarios, sample sizes were calculated for each test. The empirical type I error and power for the corresponding sample size were simulated based on 100,000 simulation runs.

The simulation results (Table 5.2) can be summarized as follows. For the OSLRT, sample sizes were always larger than that of the MOSLRT; empirical type I errors of OSLRT were less than the nominal level and empirical powers of OSLRT were over the nominal level. For the MOSLRT, sample sizes were always smaller than that of the OSLRT; empirical type I errors were close to the nominal level when $S_0(x) < 0.5$ but were slightly larger than the nominal level when $S_0(x) \geq 0.5$ and the empirical powers were close to the nominal level. Thus, the MOSLRT is a more powerful test and provided more accurate sample size estimation than that of the OSLRT.

5.3.4 Sample Size vs. Length of Follow-up

To explore the relationship between length of follow-up and sample size/number of events for the OSLRT and MOSLRT, we fixed study power at 80% and calculated sample size and number of events under exponential distribution by varying length of follow-up from 1 month to 16 months with other design parameters fixed as follows: 5-month survival probability $S_0(5) = 0.3$ for the reference group; hazard ratio $\delta = 0.7$; uniform accrual with accrual period $t_a = 10$; one-sided type I error rate $\alpha = 0.05$ and power of 80% (Table 5.3). First, the OSLRT required larger sample size and number of events than that of MOSLRT. Second, sample size decreased as the length of follow-up

TABLE 5.2: Sample size and simulated empirical type I error rate and power for OSLRT L_1 and MOSLRT L_2 under the exponential model for various designs with nominal type I error of 0.05 and power of 80%.

Design			Sample size		Type I error		Power	
(t_a, t_f, x)	S_0	S_1	n_{L_1}	n_{L_2}	L_1	L_2	L_1	L_2
(1, 1, 1)	.2	.35	45	40	.041	.051	.823	.809
	.3	.45	54	48	.041	.052	.822	.807
	.5	.65	55	48	.043	.054	.818	.802
	.6	.75	49	41	.041	.055	.819	.797
	.7	.80	90	78	.043	.056	.810	.794
(3, 1, 1)	.2	.35	42	37	.039	.050	.831	.814
	.3	.45	47	42	.041	.050	.821	.810
	.5	.65	43	38	.041	.052	.820	.809
	.6	.75	36	31	.040	.053	.824	.806
	.7	.80	63	55	.043	.055	.812	.799
(3, 2, 2)	.2	.35	44	39	.040	.051	.827	.811
	.3	.45	51	45	.042	.051	.822	.807
	.5	.65	50	44	.042	.054	.816	.802
	.6	.75	44	37	.040	.054	.820	.797
	.7	.80	80	70	.043	.056	.811	.798
(3, 3, 2)	.2	.35	41	37	.040	.049	.825	.814
	.3	.45	47	42	.042	.049	.824	.810
	.5	.65	44	38	.041	.053	.821	.802
	.6	.75	37	32	.041	.053	.819	.806
	.7	.80	66	58	.043	.054	.810	.800

TABLE 5.3: Sample size vs. length of follow-up: sample size (n) and number of events (d) were calculated for OSLRT and MOSLRT under the exponential model for various length of follow-up (t_f), with nominal type I error rate 0.05 and power of 80%. The empirical powers were simulated with sample size $n = 60$.

Follow-up	OSLRT		MOSLRT		OSLRT	MOSLRT
t_f	d	n	d	n	Power	Power
1	49	82	44	74	0.695	0.730
2	49	75	44	68	0.734	0.766
3	50	70	45	63	0.761	0.788
4	50	66	45	60	0.780	0.805
5	51	64	46	58	0.798	0.821
6	51	62	46	56	0.809	0.831
7	51	60	46	54	0.817	0.840
8	52	59	47	53	0.824	0.845
9	52	58	47	53	0.832	0.851
10	52	58	47	52	0.835	0.855
11	53	57	48	52	0.837	0.859
12	53	57	48	51	0.841	0.858
13	53	56	48	51	0.845	0.860
14	53	56	48	51	0.848	0.863
15	54	56	49	50	0.850	0.864
16	54	56	49	50	0.850	0.865

increased (Figure 5.2). However, the number of events increased as the length of follow-up increased (Figure 5.2) for both MOSLRT and OSLRT. Thus, as the length of follow-up increased, the trial required less sample size but more events. To explore the relationship between study power and length of follow-up, we fixed total sample size $n = 60$ and other design parameters were set the same as above. The empirical powers were simulated with 10,000 simulation runs by varying length of follow-up from 1 month up to 16 months (Table 5.3). The results showed that the empirical power increased as the length of follow-up increased for both MOSLRT and OSLRT and OSLRT had lower power than that of MOSLRT (Figure 5.4).

5.3.5   R Code

```
######### Sample Size Calculation for MOSLRT under Various Distributions ######
### shape is the Weibull shape parameter; S0 and S1 are survival probability###
### at fixed time point x under null and alternative; alpha and beta are    ###
### type I and type II errors; ta and tf are the accrual and follow-up time ###
### dist option is to specify one of four parametric survival distributions ###
### Note: This sample size calculation assumed that treatment distribution  ###
### has the same type distribution as reference group; For example, if the  ###
```

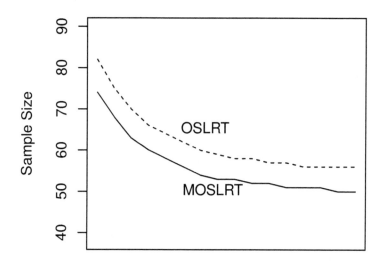

FIGURE 5.2: Sample size vs. length of follow-up for OSLRT and MOSLRT.

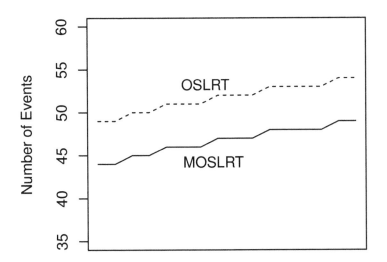

Length of follow–up (months)

FIGURE 5.3: Number of events vs. length of follow-up for OSLRT and MOSLRT.

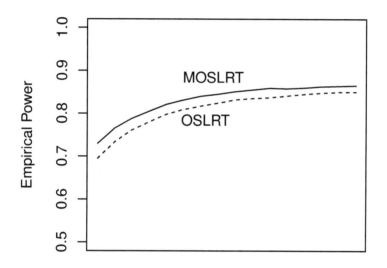

FIGURE 5.4: Empirical power vs. length of follow-up for OSLRT and MOSLRT.

```
### reference group is Weibull, then treatment group is Weibull too with   ###
### same shape parameter but different scale parameter. Same assumption is  ###
### made for the log-normal, gammma and log-logistic distribution.          ###
################################################################################
Size=function(shape,S0,S1,x,ta,tf,dist,alpha,beta)
{tau=ta+tf
 if (dist=="WB"){
  s=function(a,b,u){1-pweibull(u,a,b)}
  f=function(a,b,u){dweibull(u,a,b)}
  h=function(a,b,u){f(a,b,u)/s(a,b,u)}
  H=function(a,b,u){-log(s(a,b,u))}
  scale0=x/(-log(S0))^(1/shape)
  scale1=x/(-log(S1))^(1/shape)}

 if (dist=="LN"){
  s=function(a,b,u){1-plnorm(u,b,a)}
  f=function(a,b,u){dlnorm(u,b,a)}
  h=function(a,b,u){f(a,b,u)/s(a,b,u)}
  H=function(a,b,u){-log(s(a,b,u))}
  scale0=log(x)-shape*qnorm(1-S0)
  scale1=log(x)-shape*qnorm(1-S1)}

 if (dist=="LG"){
  s=function(a,b,u){1/(1+(u/b)^a)}
  f=function(a,b,u){(a/b)*(u/b)^(a-1)/(1+(u/b)^a)^2}
  h=function(a,b,u){f(a,b,u)/s(a,b,u)}
  H=function(a,b,u){-log(s(a,b,u))}
  scale0=x/(1/S0-1)^(1/shape)
  scale1=x/(1/S1-1)^(1/shape)}

 if (dist=="GM"){
  s=function(a,b,u){1-pgamma(u,a,b)} ## shape=a; scale=b
  f=function(a,b,u){dgamma(u,a,b)}
  h=function(a,b,u){f(a,b,u)/s(a,b,u)}
  H=function(a,b,u){-log(s(a,b,u))}
  root0=function(t){s(shape,t,x)-S0}
  scale0=uniroot(root0,c(0,10))$root
  root1=function(t){s(shape,t,x)-S1}
  scale1=uniroot(root1,c(0,10))$root}

 G=function(t){1-punif(t, tf, tau)}
 f0=function(t){s(shape,scale1,t)*h(shape,scale0,t)*G(t)}
 f1=function(t){s(shape,scale1,t)*h(shape,scale1,t)*G(t)}
 f2=function(t){s(shape,scale1,t)*h(shape,scale0,t)*H(shape,scale0,t)*G(t)}
 f3=function(t){s(shape,scale1,t)*h(shape,scale1,t)*H(shape,scale0,t)*G(t)}

 z0=qnorm(1-alpha)
 z1=qnorm(1-beta)
 v0=integrate(f0, 0, tau)$value
 v1=integrate(f1, 0, tau)$value
 v00=integrate(f2, 0, tau)$value
 v01=integrate(f3, 0, tau)$value
 n=(sqrt((v0+v1)/2)*z0+sqrt(v1+2*v0*v1-v0^2-v1^2+2*v00-2*v01)*z1)^2/(v1-v0)^2
 ans=ceiling(n)
 return(ans)
}
Size(shape=1.32,S0=0.412,S1=0.580,x=1,ta=2,tf=1,dist="WB",alpha=0.1,beta=0.1)
```

```
########## Sample Size Calculation for the MOSLRT Under PH Assumption #########
### shape is the Weibull shape parameter; S0 is survival probability at     ###
### at fixed time point $x$ under null; alpha and beta are  type I and      ###
### type II errors; delta is hazard ratio of treatment vs reference group   ###
### ta and tf are the accrual and follow-up time; dist option is to specify ###
### one of four parametric survival distribution: WB, LN, GM and LG dist.   ###
################################################################################
Size=function(shape,S0,x,delta,ta,tf,alpha,beta,dist)
{ tau=ta+tf
  scale1=delta
  if (dist=="WB"){
    scale0=x/(-log(S0))^(1/shape)
    s0=function(u){1-pweibull(u,shape,scale0)}
    f0=function(u){dweibull(u,shape,scale0)}
    h0=function(u){f0(u)/s0(u)}
    H0=function(u){-log(s0(u))}
    s=function(b,u) {s0(u)^b}
    h=function(b,u){b*f0(u)/s0(u)}
    H=function(b,u){-b*log(s0(u))}
  }

  if (dist=="LN"){
    scale0=log(x)-shape*qnorm(1-S0)
    s0=function(u){1-plnorm(u,scale0,shape)}
    f0=function(u){dlnorm(u,scale0,shape)}
    h0=function(u){f0(u)/s0(u)}
    H0=function(u){-log(s0(u))}
    s=function(b,u) {s0(u)^b}
    h=function(b,u){b*f0(u)/s0(u)}
    H=function(b,u){-b*log(s0(u))}
  }

 if (dist=="LG"){
  scale0=x/(1/S0-1)^(1/shape)
  s0=function(u){1/(1+(u/scale0)^shape)}
  f0=function(u){(shape/scale0)*(u/scale0)^(shape-1)/(1+(u/scale0)^shape)^2}
  h0=function(u){f0(u)/s0(u)}
  H0=function(u){-log(s0(u))}
  s=function(b,u) {s0(u)^b}
  h=function(b,u){b*f0(u)/s0(u)}
  H=function(b,u){-b*log(s0(u))}
 }

 if (dist=="GM"){
  s1=function(shape,b,u){1-pgamma(u,shape,b)}
  root0=function(t){s1(shape,t,x)-S0}
  scale0=uniroot(root0,c(0,10))$root
  s0=function(u){1-pgamma(u,shape,scale0)}
  f0=function(u){dgamma(u,shape,scale0)}
  h0=function(u){f0(u)/s0(u)}
  H0=function(u){-log(s0(u))}
  s=function(b,u) {s0(u)^b}
  h=function(b,u){b*f0(u)/s0(u)}
  H=function(b,u){-b*log(s0(u))}
 }
```

```
G=function(t){1-punif(t, tf, tau)}
g0=function(t){s(scale1,t)*h0(t)*G(t)}
g1=function(t){delta*s(scale1,t)*h0(t)*G(t)}
g2=function(t){s(scale1,t)*h0(t)*H0(t)*G(t)}
g3=function(t){delta*s(scale1,t)*h0(t)*H0(t)*G(t)}

z0=qnorm(1-alpha)
z1=qnorm(1-beta)
v0=integrate(g0, 0, tau)$value
v1=integrate(g1, 0, tau)$value
v00=integrate(g2, 0, tau)$value
v01=integrate(g3, 0, tau)$value
n=(sqrt((v0+v1)/2)*z0+sqrt(v1+2*v0*v1-v0^2-v1^2+2*v00-2*v01)*z1)^2/(v1-v0)^2
ans=ceiling(n)
return(ans)
}
Size(shape=1,S0=0.55,x=4,delta=0.60,ta=2,tf=4,alpha=0.05,beta=0.2,dist="WB")
[1] 52
Size(shape=1,S0=0.55,x=4,delta=0.60,ta=2,tf=4,alpha=0.05,beta=0.2,dist="LN")
[1] 50
Size(shape=1,S0=0.55,x=4,delta=0.60,ta=2,tf=4,alpha=0.05,beta=0.2,dist="GM")
[1] 52
Size(shape=1,S0=0.55,x=4,delta=0.60,ta=2,tf=4,alpha=0.05,beta=0.2,dist="LG")
[1] 54
```

5.4 Transformed OSLRT under PH Model

In previous sections, sample size calculations were done under two key assumptions: by knowing the accrual distribution and censoring distribution. However, in practice, it is extremely difficult to make the assumption for the accrual distribution due to uncontrollable accrual rate. On the other hand, the distribution of censoring time is often unknown. In this section, under a proportional hazards model, we proposed two-transformed test statistics which correct skewness of the original OSLRT statistics in an alternative way via transformation. Furthermore, we derived two number-of-events formulae which provide robust designs for the single-arm phase II survival trials evaluating survival distributions.

5.4.1 Transformed OSLRT

Let $\Lambda_0(x)$ and $S_0(x)$ be the known cumulative hazard function and survival function for the reference group, respectively, and let $\Lambda(x)$ and $S(x)$ be the unknown cumulative hazard and survival functions for the treatment group. The study considers testing the following hypothesis of interest

$$H_0 : S(x) \leq S_0(x) \quad \text{vs} \quad H_1 : S(x) > S_0(x), \quad \text{for all } x > 0.$$

Here, we will concentrate on a proportional hazards alternative, that is by assuming proportional hazards $\Lambda(x) = \delta\Lambda_0(x)$, where δ (< 1) is the hazard ratio. Thus, the hypothesis testing (5.4.1) is equivalent to testing the hazard ratio δ as follows:

$$H_0 : \delta \geq 1 \quad \text{vs} \quad H_1 : \delta < 1$$

Suppose during the accrual phase of the trial, n patients are enrolled in the study. Assuming that $\{T_i, i = 1, \ldots, n\}$ are independent failure times and $\{C_i, i = 1, \ldots, n\}$ are independent censoring times, and $\{T_i\}$ and $\{C_i\}$ are independent, then, the observed failure times and failure indicator are

$$X_i = \min(T_i, C_i) \quad \text{and} \quad \Delta_i = I(T_i \leq C_i), \quad i = 1, \cdots, n,$$

where $I(\cdot)$ is an indicator function. Let $\tilde{T}_i = \Lambda_0(T_i) = -\frac{1}{\delta}\log S(T_i)$, then $\tilde{T}_i$ follows an exponential distribution with rate parameter δ. Let $\tilde{C}_i = \Lambda_0(C_i)$, then $\Lambda_0(X_i) = \min(\tilde{T}_i, \tilde{C}_i)$. Thus, the observed data are $\{\Lambda_0(X_i), \Delta_i, i = 1, \ldots, n\}$, where $\Delta_i = I(T_i \leq C_i) = I(\tilde{T}_i \leq \tilde{C}_i)$. The likelihood function of δ given data is given by

$$L(\delta) = \prod_{i=1}^{n} \delta^{\Delta_i} e^{-\delta\Lambda_0(X_i)} = \delta^O e^{-\delta E}$$

where $O = \sum_{i=1}^{n} \Delta_i$ and $E = \sum_{i=1}^{n} \Lambda_0(X_i)$, and the log-likelihood function is

$$\ell(\delta) = O \log \delta - \delta E.$$

The first and second derivatives of the log-likelihood function are

$$\ell'(\delta) = \frac{O}{\delta} - E \quad \text{and} \quad \ell''(\delta) = -\frac{O}{\delta^2}$$

Therefore the maximum likelihood estimate of the parameter δ is given by

$$\hat{\delta} = \frac{O}{E}.$$

and its variance estimate can be obtained from the inverse of Fisher information $j^{-1}(\hat{\delta}) = -\{\ell''(\hat{\delta})\}^{-1} = O/E^2$.

To develop transformed OSLRT, we take a log transformation of $\hat{\delta}$ (see Section 2.3 of Chapter 2). Using the delta method, an approximate variance of $\log(\hat{\delta})$ is $1/O$, Thus, the standardized Wald statistic of the $\log(\hat{\delta})$ is given

$$Z_1 = \sqrt{O}\{\log(E) - \log(O)\}, \tag{5.10}$$

which is approximately standard normal distributed under the null hypothesis. Another transformed OSLRT is obtained by taking a quadratic transformation (see Section 2.4 of Chapter 2) of the $\hat{\delta}$ resulting the standardized Wald statistic given as follows:

$$Z_2 = 3O^{1/6}\{E^{1/3} - O^{1/3}\}, \tag{5.11}$$

which is approximately standard normal distributed under the null hypothesis.

Remark: Let $\theta = -\log(\delta)$ be the negative log hazard ratio, then the log-likelihood function can be written as

$$\ell(\theta) = -O\theta - e^{-\theta}E.$$

The first and second derivatives of the log-likelihood function are

$$\ell'(\theta) = -O + e^{-\theta}E \quad \text{and} \quad \ell''(\theta) = -e^{-\theta}E,$$

and Fisher information $j(\theta) = e^{-\theta}E$. Thus, the score test is given by

$$\frac{\ell'(0)}{\sqrt{j(0)}} = \frac{E - O}{\sqrt{E}}$$

which is the OSLRT.

5.4.2 Number of Events Formulae

Two transformed test statistics Z_1 and Z_2 have been discussed in Chapter 2 under the Weibull model. The number of events formulae derived in Chapter 2 can be applied here since the transformed time-to-event data follows an exponential distribution with rate δ. Therefore, from the results of Sections 2.3 and 2.4 of Chapter 2, given type I error rate α, power $1 - \beta$ and hazard ratio δ at the alternative hypothesis, the total number of events for the Z_1 test is given by (see Section 2.3 of Chapter 2)

$$d = \frac{(z_{1-\alpha} + z_{1-\beta})^2}{[\log(\delta)]^2},$$

and the total number of events required for the test statistic Z_2 can be calculated by (see Section 2.4 of Chapter 2)

$$d = \frac{(z_{1-\alpha} + z_{1-\beta})^2}{9(\delta^{-1/3} - 1)^2}.$$

Thus, under the proportional hazards model, the study design is determined by the number of events and makes no assumption for the censoring and accrual distributions which adds great flexibility and robustness to the single-arm phase II survival trial design.

5.4.3 Comparison and R Code

To explore the relationship between OSLRT, MOSLRT and two transformed OSLRT Z_1 and Z_2, we calculate the number of events and sample size under the Weibull distribution with shape parameter to be 0.5, 1, 2. We assume that 1-year survival probability is $S_0(1) = 0.3, 0.5$ for the reference group and the trial is to detect a hazard ratio $\delta = 0.7$ of the treatment vs. reference. Sample size is calculated under the uniform accrual with accrual period $t_a = 1, 2$ and follow-up time $t_f = 1, 2$ and no loss to follow-up. With one-sided type I error rate $\alpha = 0.05$ and power of 80%, the number of events and sample size required for each test are presented in Table 5.4. We have the following observations. First, the number of events and sample size for the OSLRT and MOSLRT depend on the underlying survival distribution (with different shape parameter), and accrual and censoring distributions. However, the number of events for two transformed OSLRT Z_1 and Z_2 does not depend on the underlying survival distribution as well as the accrual and censoring distributions. Second, the number of events and sample size for Z_1 test are close to that of the OSLRT and the number of events and sample size for Z_2 test are close to that of the MOSLRT. That is, the Z_1 test performs more like the OSLRT and the Z_2 test performs more like the MOSLRT. We conducted 10,000 simulations under the exponential model with number of events 49 and 44 for the Z_1 test and Z_2 test, respectively. The simulated empirical type I error and power were 0.045 and 80.4% for Z_1 test and 0.053 and 79.4% for Z_2 test. Thus, both transformed OSLRT Z_1 and Z_2 tests performed well and the Z_2 test is more efficient and requires fewer events than Z_1 test.

TABLE 5.4: Number of events (d) and sample size (n) were calculated for the OSLRT, MOSLRT and two transformed OSLRT Z_1 and Z_2 under the Weibull distribution PH model with type I error rate $\alpha = 0.05$ and power 80%.

Design parameters					OSLRT		MOSLRT		Z_1		Z_2	
Shape	$S_0(1)$	δ	t_a	t_f	d	n	d	n	d	n	d	n
.5	.3	.7	1	1	48	75	44	68	49	76	44	68
	.3	.7	2	1	49	71	44	64	49	71	44	63
	.3	.7	1	2	49	67	45	60	49	67	44	59
	.3	.7	2	2	50	65	45	59	49	64	44	57
	.5	.7	1	1	46	104	42	93	49	110	44	97
	.5	.7	2	1	47	95	42	85	49	100	44	88
	.5	.7	1	2	47	88	42	79	49	91	44	81
	.5	.7	2	2	48	84	43	75	49	86	44	77
1	.3	.7	1	1	49	69	44	62	49	69	44	61
	.3	.7	2	1	51	64	46	58	49	62	44	55
	.3	.7	1	2	52	59	47	53	49	56	44	50
	.3	.7	2	2	52	58	47	52	49	54	44	48
	.5	.7	1	1	47	92	42	82	49	95	44	85
	.5	.7	2	1	48	79	43	71	49	81	44	72
	.5	.7	1	2	49	70	44	63	49	70	44	62
	.5	.7	2	2	50	66	45	59	49	65	44	57
2	.3	.7	1	1	51	62	46	56	49	60	44	53
	.3	.7	2	1	53	59	48	53	49	54	44	48
	.3	.7	1	2	55	55	50	50	49	50	44	44
	.3	.7	2	2	55	55	50	50	49	49	44	44
	.5	.7	1	1	49	75	44	68	49	75	44	67
	.5	.7	2	1	52	65	47	58	49	62	44	55
	.5	.7	1	2	53	56	48	51	49	52	44	46
	.5	.7	2	2	54	56	49	51	49	51	44	45

```
Size=function(kappa,S0,delta,x,ta,tf,alpha,beta)
{tau=ta+tf
 lambda0=-log(S0)/x^kappa
 lambda1=delta*lambda0
 S0=function(t){exp(-lambda0*t^kappa)}
 h0=function(t){kappa*lambda0*t^(kappa-1)}
 H0=function(t){lambda0*t^kappa}
 S1=function(t){exp(-lambda1*t^kappa)}
 h1=function(t){kappa*lambda1*t^(kappa-1)}
 H1=function(t){lambda1*t^kappa}
 G=function(t){1-punif(t, tf, tau)}
 f0=function(t){S1(t)*h0(t)*G(t)}
 f1=function(t){S1(t)*h1(t)*G(t)}
 f2=function(t){S1(t)*h0(t)*H0(t)*G(t)}
 f3=function(t){S1(t)*h1(t)*H0(t)*G(t)}
 z0=qnorm(1-alpha)
 z1=qnorm(1-beta)
```

```
v0=integrate(f0, 0, tau)$value
v1=integrate(f1, 0, tau)$value
v00=integrate(f2, 0, tau)$value
v01=integrate(f3, 0, tau)$value
p=v1
n1=(sqrt(v0)*z0+sqrt(v1+2*v0*v1-v0^2-v1^2+2*v00-2*v01)*z1)^2/(v1-v0)^2
d1=ceiling(n1*p)
n2=(sqrt((v0+v1)/2)*z0+sqrt(v1+2*v0*v1-v0^2-v1^2+2*v00-2*v01)*z1)^2/(v1-v0)^2
d2=ceiling(n2*p)
d3=(z0+z1)^2/(log(delta)^2)
n3=ceiling(d3/p)
d4=(z0+z1)^2/((delta^(-1/3)-1)^2*9)
n4=ceiling(d4/p)
n1=ceiling(n1)
n2=ceiling(n2)
d3=ceiling(d3)
d4=ceiling(d4)
ans=c(d1,n1,d2,n2,d3,n3,d4,n4)
return(ans)
}
Size(kappa=0.5,S0=0.3,delta=0.7,x=1,ta=1,tf=1,alpha=0.05,beta=0.2)
48    75    44    68    49    76    44    68
```

5.5   General Hypothesis Testing

In some cases, the experimental treatment is less toxic, or less expensive than
the standard of care. In this case, we may want to show that an experimental
treatment is equal or not inferior to the reference. For designing such trials,
we can use a generalized OSLRT which is discussed by Sun et al. (2011). Let
$\lambda_0(x)$ and $S_0(x)$ be the known baseline hazard function and survival function,
respectively, under the null hypothesis, and let $\lambda(x)$ and $S(x)$ be the unknown
hazard and survival functions for the treatment group, and $\delta = \lambda(x)/\lambda_0(x)$
be the hazard ratio which is a constant. We consider the following general
hypothesis:

$$H_0 : \delta \geq \delta_0 \quad \text{vs} \quad H_1 : \delta < \delta_0, \tag{5.12}$$

for some prefixed $\delta_0(\geq 1)$. The study is powered at the alternative hypothesis
$H_1 : \delta = \delta_1$ for some $\delta_1 < \delta_0$. The experimental treatment is not worthy
of further evaluation if its survival distribution falls significantly below the
survival curve $[S_0(x)]^{\delta_0}$ and the experimental treatment is promising if its
survival distribution locates significantly above the survival curve $[S_0(x)]^{\delta_0}$
and concludes to favor the alternative survival distribution $[S_0(x)]^{\delta_1}$.

5.5.1 Test Statistics

To test hypothesis (5.12), we use the following generalized OSLRT

$$L = \frac{\delta_0 E - O}{\sqrt{\delta_0 E}}.$$ (5.13)

When $\delta_0 = 1$ which is reduced to the OSLRT. Similar to the OSLRT, L is asymptotically standard normal distributed under the null hypothesis. Hence, we reject null hypothesis H_0 with one-sided type I error rate α if $L > z_{1-\alpha}$, where $z_{1-\alpha}$ is the $100(1-\alpha)$ percentile of the standard normal distribution.

5.5.2 Sample Size Formula

Same derivation as the OSLRT, the sample size formula for the generalized OSLRT is give by

$$n = \frac{(\sigma_0 z_{1-\alpha} + \sigma z_{1-\beta})^2}{\omega^2},$$ (5.14)

where $\omega = \sigma_0^2 - \sigma_1^2$ and $\sigma^2 = v_1 - v_1^2 + 2v_{00} - v_0^2 - 2v_{01} + 2v_0 v_1$, with v_0, v_1, v_{00} and v_{01} given as follows:

$$v_0 = \sigma_0^2 = \delta_0 \int_0^\infty G(x) S_1(x) d\Lambda_0(x),$$

$$v_1 = \sigma_1^2 = \delta_1 \int_0^\infty G(x) S_1(x) d\Lambda_0(x),$$

$$v_{00} = \delta_0^2 \int_0^\infty G(x) S_1(x) \Lambda_0(x) d\Lambda_0(x),$$

$$v_{01} = \delta_0 \delta_1 \int_0^\infty G(x) S_1(x) \Lambda_0(x) d\Lambda_0(x).$$

where $S_1(x) = [S_0(x)]^{\delta_1}$.

Remark 3: The choice $\delta_0 = 1$ and $\delta_1 < 1$ represents the superiority hypothesis of proving that the survival under the experimental treatment arm is superior to the reference (under the null hypothesis). The choice $\delta_0 > 1$ and $\delta_1 \leq 1$ represents the non-inferiority of the experimental treatment to the reference where δ_0 is determined by the survival distribution $S_0(x)$ and non-inferiority margin.

5.5.3 Simulation

To study the performance of the generalized OSLRT under a general hypothesis setting, we calculated sample sizes under both superior and non-inferior hypotheses scenarios. Under the Weibull distribution, the shape parameter

TABLE 5.5: Sample size (n), simulated empirical type I error rate ($\hat{\alpha}$), and empirical power (EP) for the OSLRT based on 100,000 simulation runs for the Weibull distributions with nominal type I error of 0.05 and power of 80%. The censoring distribution is uniform over $[t_f, t_a + t_f]$, where $t_a = 1$ and $t_f = 2$.

Design			$\kappa = 0.5$			$\kappa=1$			$\kappa = 2$		
$S_0(1)$	δ_0	δ_1	n	α	EP	n	$\hat{\alpha}$	EP	n	$\hat{\alpha}$	EP
.2	1	.7	61	.053	.818	56	.051	.820	55	.049	.825
.3	1	.7	67	.053	.817	59	.052	.821	55	.051	.817
.4	1	.7	76	.054	.818	63	.054	.819	55	.052	.818
.5	1	.7	88	.057	.812	70	.058	.817	56	.053	.817
.6	1	.7	107	.056	.811	81	.055	.816	59	.055	.817
.2	1.3	1	103	.053	.815	99	.051	.816	98	.049	.818
.3	1.3	1	109	.053	.814	101	.052	.817	98	.051	.817
.4	1.3	1	119	.054	.812	105	.054	.815	99	.052	.818
.5	1.3	1	135	.057	.812	113	.058	.815	99	.053	.817
.6	1.3	1	159	.056	.810	126	.055	.814	101	.055	.817
.2	1.3	.9	55	.053	.824	52	.051	.822	52	.049	.825
.3	1.3	.9	58	.053	.820	54	.052	.828	52	.051	.826
.4	1.3	.9	64	.054	.819	56	.054	.822	52	.052	.825
.5	1.3	.9	72	.057	.816	60	.058	.819	52	.053	.822
.6	1.3	.9	85	.056	.810	67	.055	.815	54	.055	.828

was set to 0.5, 1, and 2 to reflect the different types of hazard function; survival probabilities under the null $S_0(1)$ was set to be 0.2 to 0.6 by 0.1. We assumed that subjects were recruited with a uniform distribution over the accrual period t_a of 1 year and followed for a period t_f of 2 years. The censoring distribution was a uniform distribution on the interval $[t_f, t_a + t_f]$. The required sample sizes for each design scenario were calculated. The empirical type I error and the power of the corresponding design were simulated based on 100,000 runs (Table 5.5). The simulation results showed that both the empirical type I error rate and empirical power were slightly over the nominal levels. Thus, the generalized OSLRT was slightly liberal and sample sizes were slightly overestimated.

Example 5 *Trial for Breast Cancer*

In a phase III cancer clinical trial, it is shown that premenopausal women with axillary node-positive breast cancer following a surgical resection and treated by a new chemotherapy regimen CEF (colony stimulating factor) had longer medium duration of the relapse-free survival (RFS) in comparison with those treated by a standard regimen CMF (Levine et al., 1998). In order to further improve CEF, a phase I/II clinical trial was conducted to identify the dose for a new combination CEF with granulocyte colony stimulating factor (G-CSF) support in premenopausal women with four or more axillary node-

positive operable breast cancer and evaluate its activity. A total of 63 patients were registered in the study. After the maximum tolerated dose (MTD) had been established, all registered patients were followed for the assessment of RFS (Findlay et al., 2007).

From historical data, the estimated 4-year RFS for patients with four or more positive nodes and treated with classical CEF regimen was around 55%, that is $S_0(4) = 0.55$. We would consider the modified CEF as not worthy of further evaluation if its 4-year RFS is 50% or lower (corresponding to a 5% non-inferiority margin) or $[S_0(4)]^{\delta_0} = 0.5$ and promising if its 4-year RFS is 70% or higher or $S_1(4) = [S_0(4)]^{\delta_1} = 0.7$. If we assume an exponential distribution for RFS, we can find out this is equivalent to assuming $\delta_0 = \log(0.5)/\log(0.55) = 1.16$ and $\delta_1 = \log(0.70)/\log(0.55) = 0.60$. Assume uniform accrual with accrual period of $t_a = 2$ years and follow-up duration $t_f = 4$ years (from Sun et al., 2003). Using the formula (5.14), the required sample size is $n = 35$ with 80% power and 5% type I error.

If annual accrual rate is 18 patients, using R code 'Duration', the calculated the study duration is 1.9324 years and required sample size is $1.9324 \times 18 \simeq 35$ patients which is same as the sample size calculated using formula (5.14).

```
########## Sample Size Calculation for the Generalized OSLRT ##################
### shape is the Weibull shape parameter; S0 is survival probability at    ###
### at fixed time point $x$ under null; alpha and beta are  type I and      ###
### type II errors; gamma0 and gamma1 are the parameter for the general    ###
### hypothesis testing; ta and tf are the accrual and follow-up time;      ###
### dist option is to specify one of four parametric survival distributions ###
##############################################################################
Size=function(shape,S0,x,delta0,delta1,ta,tf,alpha,beta,dist)
{ tau=ta+tf
  scale1=delta1
  if (dist=="WB"){
    scale0=x/(-log(S0))^(1/shape)
    s0=function(u){1-pweibull(u,shape,scale0)}
    f0=function(u){dweibull(u,shape,scale0)}
    h0=function(u){f0(u)/s0(u)}
    H0=function(u){-log(s0(u))}
    s=function(b,u) {s0(u)^b}
    h=function(b,u){b*f0(u)/s0(u)}
    H=function(b,u){-b*log(s0(u))}
  }

  if (dist=="LN"){
    scale0=log(x)-shape*qnorm(1-S0)
    s0=function(u){1-plnorm(u,scale0,shape)}
    f0=function(u){dlnorm(u,scale0,shape)}
    h0=function(u){f0(u)/s0(u)}
    H0=function(u){-log(s0(u))}
    s=function(b,u) {s0(u)^b}
    h=function(b,u){b*f0(u)/s0(u)}
    H=function(b,u){-b*log(s0(u))}
  }

  if (dist=="LG"){
  a=shape
```

```
    b=scale0=x/(1/S0-1)^(1/shape)
    s0=function(u){1/(1+(u/b)^a)}
    f0=function(u){(a/b)*(u/b)^(a-1)/(1+(u/b)^a)^2}
    h0=function(u){f0(u)/s0(u)}
    H0=function(u){-log(s0(u))}
    s=function(b,u) {s0(u)^b}
    h=function(b,u){b*f0(u)/s0(u)}
    H=function(b,u){-b*log(s0(u))}
    }

    if (dist=="GM"){
    s1=function(shape,b,u){1-pgamma(u,shape,b)}
    root0=function(t){s1(shape,t,x)-S0}
    scale0=uniroot(root0,c(0,10))$root
    s0=function(u){1-pgamma(u,shape,scale0)}
    f0=function(u){dgamma(u,shape,scale0)}
    h0=function(u){f0(u)/s0(u)}
    H0=function(u){-log(s0(u))}
    s=function(b,u) {s0(u)^b}
    h=function(b,u){b*f0(u)/s0(u)}
    H=function(b,u){-b*log(s0(u))}
    }

    G=function(t){1-punif(t, tf, tau)}
    g0=function(t){delta0*s(scale1,t)*h0(t)*G(t)}
    g1=function(t){delta1*s(scale1,t)*h0(t)*G(t)}
    g2=function(t){delta0^2*s(scale1,t)*h0(t)*H0(t)*G(t)}
    g3=function(t){delta0*delta1*s(scale1,t)*h0(t)*H0(t)*G(t)}

    z0=qnorm(1-alpha)
    z1=qnorm(1-beta)
    v0=integrate(g0, 0, tau)$value
    v1=integrate(g1, 0, tau)$value
    v00=integrate(g2, 0, tau)$value
    v01=integrate(g3, 0, tau)$value
    n=(sqrt(v0)*z0+sqrt(v1+2*v0*v1-v0^2-v1^2+2*v00-2*v01)*z1)^2/(v1-v0)^2
    ans=ceiling(n)
    return(ans)
    }
Size(shape=1,S0=0.55,x=4,delta0=1.16,delta1=0.60,ta=2,tf=4,alpha=0.05,
beta=0.2,dist="WB")
[1] 35

########## Study Duration Calculation for the Generalized OSLRT ##############
### shape is the Weibull shape parameter; S0 is survival probability at   ###
### at fixed time point $x$ under null; alpha and beta are  type I and     ###
### type II errors; gamma0 and gamma1 are the parameter for the general    ###
### hypothesis testing; rate is accural rate per unit time; tf are the     ###
### follow-up time; dist option is to specify one of the four parametric   ###
### survival distributions.                                                ###
#############################################################################
Duration=function(shape,S0,x,delta0,delta1,rate,tf,alpha,beta,dist)
{
  if (dist=="WB"){
    scale0=x/(-log(S0))^(1/shape)
    s0=function(u){1-pweibull(u,shape,scale0)}
    f0=function(u){dweibull(u,shape,scale0)}
```

```
   h0=function(u){f0(u)/s0(u)}
   H0=function(u){-log(s0(u))}
   s=function(b,u) {s0(u)^b}
   h=function(b,u){b*f0(u)/s0(u)}
   H=function(b,u){-b*log(s0(u))}
 }

 if (dist=="LN"){
   scale0=log(x)-shape*qnorm(1-S0)
   s0=function(u){1-plnorm(u,scale0,shape)}
   f0=function(u){dlnorm(u,scale0,shape)}
   h0=function(u){f0(u)/s0(u)}
   H0=function(u){-log(s0(u))}
   s=function(b,u) {s0(u)^b}
   h=function(b,u){b*f0(u)/s0(u)}
   H=function(b,u){-b*log(s0(u))}
 }

 if (dist=="LG"){
  a=shape
  b=scale0=x/(1/S0-1)^(1/shape)
  s0=function(u){1/(1+(u/b)^a)}
  f0=function(u){(a/b)*(u/b)^(a-1)/(1+(u/b)^a)^2}
  h0=function(u){f0(u)/s0(u)}
  H0=function(u){-log(s0(u))}
  s=function(b,u) {s0(u)^b}
  h=function(b,u){b*f0(u)/s0(u)}
  H=function(b,u){-b*log(s0(u))}
 }

 if (dist=="GM"){
  s1=function(shape,b,u){1-pgamma(u,shape,b)}
  root0=function(t){s1(shape,t,x)-S0}
  scale0=uniroot(root0,c(0,10))$root
  s0=function(u){1-pgamma(u,shape,scale0)}
  f0=function(u){dgamma(u,shape,scale0)}
  h0=function(u){f0(u)/s0(u)}
  H0=function(u){-log(s0(u))}
  s=function(b,u) {s0(u)^b}
  h=function(b,u){b*f0(u)/s0(u)}
  H=function(b,u){-b*log(s0(u))}
 }
 root=function(ta){
  tau=ta+tf
  scale1=delta1
  G=function(t){1-punif(t, tf, tau)}
  g0=function(t){delta0*s(scale1,t)*h0(t)*G(t)}
  g1=function(t){delta1*s(scale1,t)*h0(t)*G(t)}
  g2=function(t){delta0^2*s(scale1,t)*h0(t)*H0(t)*G(t)}
  g3=function(t){delta0*delta1*s(scale1,t)*h0(t)*H0(t)*G(t)}

  z0=qnorm(1-alpha)
  z1=qnorm(1-beta)
  v0=integrate(g0, 0, tau)$value
  v1=integrate(g1, 0, tau)$value
  v00=integrate(g2, 0, tau)$value
  v01=integrate(g3, 0, tau)$value
```

```
  ans=rate*ta-(sqrt(v0)*z0+sqrt(v1+2*v0*v1-v0^2-v1^2+2*v00-2*v01)*z1)^2/(v1-v0)^2
  }
 ta=uniroot(root, lower=0, upper=5*tf)$root
 n=ceiling(rate*ta); ta=round(ta,4)
 ans=list(ta=ta,n=n)
 return(ans)
}
Duration(shape=1,S0=0.55,x=4,delta0=1.16,delta1=0.60,rate=18,tf=4,alpha=0.05,
beta=0.2,dist="WB")
$ta
[1] 1.9324
$n
[1] 35
```

6

Two-Stage Design Evaluating Survival Distributions

6.1 Introduction

In this chapter, an optimal two-stage design using OSLRT proposed by Kwak and Jung (2014) will be discussed. Kwak and Jung's two-stage design required each patient to be followed until an event occurs or until the end of the study, whichever occurs first. In practice, however, full follow-up information for a phase II trial may be difficult to obtain in the late period of the trial, particularly when the accrual duration is relatively longer; obtaining the status of each patient within a restricted period is more realistic (Belin et al., 2017). Therefore we will also present a two-stage design with restricted follow-up time by using OSLRT (Belin et al., 2017; Wu et al., 2019). The two-stage design using MOSLRT is also briefly discussed.

6.2 Two-Stage Design Using OSLRT with Full Follow-up

Suppose during the accrual phase of the trial, n subjects are enrolled in the study at calendar times $A_1, \cdots, A_n$, which are measured from the start of the study. Let T_i and C_i denote, respectively, the failure time and censoring time of the ith subject, with both being measured from the time of study entry of this subject. We assume that the failure time T_i is independent of the censoring time C_i and entry time A_i, and $\{(A_i, T_i, C_i); i = 1, \cdots, n\}$ are independent and identically distributed.

When the data are examined at calendar time $t \leq \tau$, we observe the time-to-failure $X_i(t) = T_i \wedge C_i \wedge (t - A_i)^+$ and failure indicator $\Delta_i(t) = I(T_i \wedge C_i \wedge (t - A_i)^+), i = 1, \cdots, n$, where A_i is uniformly distributed on $[0, t_a]$. Based on the observed data $\{X_i(t), \Delta_i(t), i = 1, \cdots, n\}$, let $N_i(t, u) = \Delta_i(t)I(X_i(t) \leq u)$ and $Y_i(t, u) = I(X_i(t) \geq u)$ be the failure and at-risk process, respectively. Consider a two-stage design with an interim analysis at calendar time t_1 for stopping futility only and final analysis at calendar time τ, corresponding sample sizes are n_1 and n for the interim analysis and final

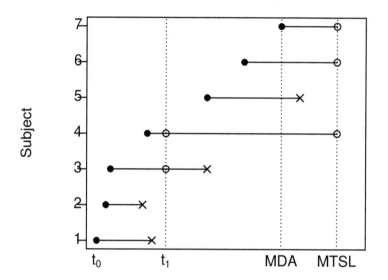

FIGURE 6.1: Two-stage survival data with unrestricted follow-up time.

analysis, respectively. We further assume no subjects lost to follow-up during the study; that is $C_i = +\infty$ for all i, thus, due to staggered entry, censoring of ith subject is caused by the administrative censoring variable $(t - A_i)^+$ at calendar time t.

6.2.1 Two-Stage Test Statistics

With full follow-up data, we will test the following hypothesis

$$H_0 : S(u) \leq S_0(u) \quad \text{vs} \quad H_1 : S(u) > S_0(u), \quad 0 < u \leq \tau, \tag{6.1}$$

where $S_0(u)$ and $S(u)$ are the survival distributions of the reference group and new treatment group, respectively. Let n_1 and n be the sample size of stage I and stage II of the trial and corresponding calendar times are t_1 and τ, we define $O_1 = \sum_{i=1}^{n_1} \Delta_i(t_1)$ as the observed number of events, and $E_1 = \sum_{i=1}^{n_1} \Lambda_0(X_i(t_1))$ as the expected number of events for stage I, and $O = \sum_{i=1}^{n} \Delta_i(\tau)$ as the observed number of events, and $E = \sum_{i=1}^{n} \Lambda_0(X_i(\tau))$

as the expected number of events up to stage II, where $\Lambda_0(x)$ is the cumulative hazard function under the null hypothesis. Let $Z_1 = (E_1 - O_1)/\sqrt{E_1}$ and $Z = (E - O)/\sqrt{E}$ be the OSLRT statistics for stage I and stage II of the trial. Using counting process notations, we have $Z_1 = U_1/\hat{\sigma}_1$ and $Z = U/\hat{\sigma}$, where

$$U_1 = n_1^{-1/2} \sum_{i=1}^{n_1} \int_0^\tau \{Y_i(t_1, u)d\Lambda_0(u) - N_i(t_1, du)\},$$

$$U = n^{-1/2} \sum_{i=1}^{n} \int_0^\tau \{Y_i(\tau, u)d\Lambda_0(u) - N_i(\tau, du)\}$$

and

$$\hat{\sigma}_1^2 = n_1^{-1} \sum_{i=1}^{n_1} \int_0^\tau Y_i(t_1, u)d\Lambda_0(u),$$

$$\hat{\sigma}^2 = n^{-1} \sum_{i=1}^{n} \int_0^\tau Y_i(\tau, u)d\Lambda_0(u).$$

Under the null hypothesis H_0, the means $E(U_1|H_0) = E(U|H_0) = 0$ and variances $\text{var}(U_1|H_1) = \nu_1$ and $\text{var}(U|H_1) = \nu$, where ν_1 and ν are given as follows:

$$\nu_1 = \int_0^\tau G(u, t_1)S_0(u)d\Lambda_0(u),$$

$$\nu = \int_0^\tau G(u, \tau)S_0(u)d\Lambda_0(u),$$

and $G(u, t_1) = P(t_1 - A_1 > u) = P(A_1 < t_1 - u)$ with A_1 uniformly distributed on $[0, t_a]$. Thus, $G(u, t_1)$ is given as follows:

$$G(u, t_1) = \begin{cases} 1 & u \le t_1 - t_a \\ (t_1 - u)/t_a & t_1 - t_a < u \le t_1 \\ 0 & u > t_1 \end{cases}$$

Under the null hypothesis H_0, by independent increment of the OSLRT, it shows $\text{cov}(U_1, U|H_0) = \text{cov}(U_1, U_1|H_0) = \text{var}(U_1|H_0) = \nu_1$. Thus, the variance matrix of (U_1, U) is

$$\tilde{\Sigma} = \begin{pmatrix} \nu_1 & \nu_1 \\ \nu_1 & \nu \end{pmatrix}.$$

Furthermore $n_1^{-1}\sum_{i=1}^{n_1} Y_i(t_1, u)$, and $n^{-1}\sum_{i=1}^{n} Y_i(t_2, u)$ uniformly converge to $S_0(u)G(u, t_1)I(u \le x)$ and $S_0(u)I(u \le x)$, respectively, then, $\hat{\sigma}_1^2 \to \nu_1$ and $\hat{\sigma}^2 \to \nu$, or $\text{Var}(Z_1|H_0) = \text{Var}(Z_1|H_0) \simeq 1$. Therefore, the correlation between Z_1 and Z is approximately given as $\rho_0 = \text{cov}(Z_1, Z|H_0) \simeq \text{cov}(U_1, U|H_0)/\sqrt{\nu_1\nu} = \sqrt{\nu_1/\nu}$. Thus, under the null, (Z_1, Z) is approximately bivariate normal distributed with mean $\mu = (0, 0)'$ and variance matrix

$$\Sigma = \begin{pmatrix} 1 & \rho_0 \\ \rho_0 & 1 \end{pmatrix}$$

Under alternative H_1, we have $E(U_1|H_1) = \sqrt{n_1}\omega_1$ and $E(U|H_1) = \sqrt{n}\omega$, where ω_1 and ω are given as follows:

$$\omega_1 = \int_0^\tau G(u, t_1)S_1(u)d\Lambda_0(u) - \int_0^\tau G(u, t_1)S_1(u)d\Lambda_1(u)$$

and

$$\omega = \int_0^\tau G(u, \tau)S_1(u)d\Lambda_0(u) - \int_0^\tau G(u, \tau)S_1(u)d\Lambda_1(u)$$

and the exact variance of U_1 under H_1 is given by $\sigma_{11}^2 = p_1 - p_1^2 + 2p_{00} - p_0^2 - 2p_{01} + 2p_0 p_1$ with p_0, p_1, p_{00} and p_{01} given as follows:

$$p_0 = \int_0^\tau G(u, t_1)S_1(u)d\Lambda_0(u),$$

$$p_1 = \int_0^\tau G(u, t_1)S_1(u)d\Lambda_1(u),$$

$$p_{00} = \int_0^\tau G(u, t_1)S_1(u)\Lambda_0(u)d\Lambda_0(u),$$

$$p_{01} = \int_0^\tau G(u, t_1)S_1(u)\Lambda_0(u)d\Lambda_1(u),$$

and the exact variance of U under H_1 is given by $\sigma_{21}^2 = v_1 - v_1^2 + 2v_{00} - v_0^2 - 2v_{01} + 2v_0 v_1$, with v_0, v_1, v_{00} and v_{01} given as follows:

$$v_0 = \int_0^\tau G(u, \tau)S_1(u)d\Lambda_0(u),$$

$$v_1 = \int_0^\tau G(u, \tau)S_1(x)d\Lambda_1(u),$$

$$v_{00} = \int_0^\tau G(u, \tau)S_1(u)\Lambda_0(u)d\Lambda_0(u),$$

$$v_{01} = \int_0^\tau G(u, \tau)S_1(u)\Lambda_0(u)d\Lambda_1(u).$$

Similarly, under the alternative hypothesis H_1, by independent increment of the OSLRT, it shows $\text{cov}(U_1, U|H_1) = \text{var}(U_1|H_1) = \sigma_{11}^2$. As the exact variance $\text{var}(U|H_1) = \sigma_{21}^2$, thus, the correlation between U_1 and U under alternative is given by $\rho_1 = \sigma_{11}/\sigma_{21}$. Furthermore, it can be shown $\hat{\sigma}_1^2 \to \sigma_{01}^2 = \int_0^\tau G(u, t_1)S_1(u)d\Lambda_0(u)$ and $\hat{\sigma}^2 \to \sigma_0^2 = \int_0^\tau S_1(u)d\Lambda_0(u)$. Therefore, under the alternative, (Z_1, Z) is approximately bivariate normal distributed with mean $\mu = (\sqrt{n_1}\omega_1/\sigma_{01}, \sqrt{n}\omega/\sigma_0)'$ and variance matrix

$$\Sigma = \begin{pmatrix} \sigma_{11}^2/\sigma_{01}^2 & \rho_1\sigma_{11}\sigma_{21}/\sigma_{01}\sigma_0 \\ \rho_1\sigma_{11}\sigma_{21}/\sigma_{01}\sigma_0 & \sigma_{21}^2/\sigma_0^2 \end{pmatrix}$$

For a two-stage design, we consider stopping for futility only at first interim analysis. Suppose that the stopping boundaries are c_1 and c for the first stage and final analysis. At the first stage we stop for futility if $Z_1 \leq c_1$. Otherwise, the trial continues to the second stage and we accept futility or efficacy according to $Z \leq c$ or $Z > c$, respectively, where the boundaries c_1 and c satisfy the following type I error rate α and power $1 - \beta$ constraints:

$$P(Z_1 > c_1, Z > c \,|H_0) = \alpha, \tag{6.2}$$
$$P(Z_1 > c_1, Z > c \,|H_1) = 1 - \beta. \tag{6.3}$$

The two-stage design parameters (n_1, c_1, n, c) are unknown. An iterative algorithm will be implemented to determine (n_1, c_1, n, c) under the following set up of the design parameters.

- Uniform accrual with a constant accrual rate r.

- Survival probability $S_0(x_0)$ at fixed time point x_0.

- Survival distribution function under the null or a shape parameter for a parametric distribution.

- Hazard ratio δ, a proportional hazard model is assumed under the alternative.

- Follow-up time t_f after last patient enrolled in the study.

6.2.2 Optimal Two-Stage Procedure

1. Given $(\alpha, \beta, r, S_0, x_0, \delta, t_f, \text{dist})$, calculate sample size n_0 and accrual period $t_a^0 = n_0/r$ required for a single-stage design, where r is a constant accrual rate; S_0 is the survival probability under the null at a fixed time point x_0; t_f is the follow-up duration specified for the trial; "dist" augment is used to specify the underlying survival distribution under the null hypothesis. In our algorithm, we implemented four parametric distributions: Weibull (WB), log-normal (LN), gamma (GM) and log-logistic (LG); With a parametric distribution, we only need to specify its shape parameter; Survival distribution under the alternative is determined by proportional hazard assumption.

2. Initial algorithm with $n_1 = n = n_0$, and $c_1 = 0.25$, and given c, calculate α_c by using following equation

$$\alpha_c = \int_c^\tau \phi(z)\Phi\left(\frac{\rho_0 z - c_1}{\sqrt{1 - \rho_0^2}}\right) dz \tag{6.4}$$

and iterate it until α_c close to the nominal type I error rate α.

3. Calculate power of a given design (n_1, c_1, n, c) using the following equation

$$\text{power} = \int_{\bar{c}}^\tau \phi(z)\Phi\left(\frac{\rho_1 z - \bar{c}_1}{\sqrt{1 - \rho_1^2}}\right) dz \tag{6.5}$$

where

$$\bar{c}_1 = \frac{\sigma_{01}}{\sigma_{11}}\left(c_1 - \frac{\sqrt{n_1}\omega_1}{\sigma_{01}}\right) \quad \text{and} \quad \bar{c} = \frac{\sigma_0}{\sigma_{21}}\left(c - \frac{\sqrt{n}\omega}{\sigma_0}\right)$$

4. If the power is smaller than $1 - \beta$, then (n_1, c_1, n, c) is left and a new triplet (n_1, c_1, n) is tested by repeating steps 2 and 3, otherwise, (n_1, c_1, n, c) is selected.

For each selected candidate design (n_1, c_1, n, c), the expected sample size under H_0 is calculated by $E(n|H_0) = n_1 + (1 - \text{PS}_0)(n - n_1)$, where PS_0 is the probability of early stopping under H_0, that is $\text{PS}_0 = \Phi(c_1)$. Thus, the optimal design is one (n_1, c_1, n, c) which minimizes $E(n|H_0)$. R function 'Optimal.KJ' implemented above optimal algorithm is given in the next subsection. 'Optimal.KJ' refers to optimal two-stage Kwak and Jung's design with full follow-up data.

Remark: Two-stage design using the MOSLRT can be developed in the same fashion and corresponding R code is given in Appendix K.

6.2.3   R Code for Two-Stage Design

Following optimal two-stage design R code 'Optimal.KJ' is modified based on the algorithm developed by Kwak and Jung (2013). The four parametric survival distributions implemented in the 'Optimal.KJ' code are the Weibull distribution (WB), log-normal (LN), gamma (GM) and log-logistic (LG) distributions. The R code is implemented under the proportional hazards model, that is $S_1(t) = [S_0(t)]^\delta$, where δ is the hazard ratio.

```
#################### Optimal.KJ Input Parameters ##########################
###   shape is the shape parameter for one the four parametric distributions;###
###   S0 is the survival probability at fixed time point x0 under the null;  ###
###   hr is the hazard ratio; rate is constant accrual rate; alpha and beta  ###
###   are the type I and II errors; dist is distribution option with 'WB' as ###
###   Weibull, 'GM' as Gamma, 'LN' as log-normal, 'LG' as log-logistic.      ###
##########################################################################
library(survival)
Optimal.KJ<-function(shape,S0,x0,hr,tf,rate,alpha,beta,dist)
{
  calculate_alpha<-function(c2, c1, rho0){
    fun1<-function(z, c1, rho0){
      f<-dnorm(z)*pnorm((rho0*z-c1)/sqrt(1-rho0^2))
      return(f)
    }
    alpha<-integrate(fun1, lower= c2, upper= Inf, c1, rho0)$value
    return(alpha)
  }

  calculate_power<-function(cb, cb1, rho1){
    fun2<-function(z, cb1, rho1){
      f<-dnorm(z)*pnorm((rho1*z-cb1)/sqrt(1-rho1^2))
```

```
      return(f)
    }
  pwr<-integrate(fun2,lower=cb,upper=Inf,cb1=cb1,rho1=rho1)$value
  return(pwr)
}

fct<-function(zi, ceps=0.0001,alphaeps=0.0001,nbmaxiter=100,dist){
  ta<-as.numeric(zi[1])
  t1<-as.numeric(zi[2])
  c1<-as.numeric(zi[3])

if (dist=="WB"){
  f0=function(t){(shape/scale0)*(t/scale0)^(shape-1)*exp(-(t/scale0)^shape)}
  s0=function(t){exp(-(t/scale0)^shape)}
  h0=function(t){(shape/scale0)*(t/scale0)^(shape-1)}
  H0=function(t){(t/scale0)^shape}
  s=function(b, t){exp(-b*(t/scale0)^shape)}
  h=function(b,t){b*h0(t)}
  H=function(b,t){b*H0(t)}
  scale0=x0/(-log(S0))^(1/shape)
  scale1=hr
}


if (dist=="LN"){
  s0=function(u){1-plnorm(u,scale0,shape)}
  f0=function(u){dlnorm(u,scale0,shape)}
  h0=function(u){f0(u)/s0(u)}
  H0=function(u){-log(s0(u))}
  s=function(b,u) {s0(u)^b}
  h=function(b,u){b*f0(u)/s0(u)}
  H=function(b,u){-b*log(s0(u))}
  scale0=log(x0)-shape*qnorm(1-S0)
  scale1=hr
}

if (dist=="LG"){
  s0=function(u){1/(1+(u/scale0)^shape)}
  f0=function(u){(shape/scale0)*(u/scale0)^(shape-1)/(1+(u/scale0)^shape)^2}
  h0=function(u){f0(u)/s0(u)}
  H0=function(u){-log(s0(u))}
  s=function(b,u) {s0(u)^b}
  h=function(b,u){b*f0(u)/s0(u)}
  H=function(b,u){-b*log(s0(u))}
  scale0=x0/(1/S0-1)^(1/shape)
  scale1=hr
}

if (dist=="GM"){
  s0=function(u){1-pgamma(u,shape,scale0)}
  f0=function(u){dgamma(u,shape,scale0)}
  h0=function(u){f0(u)/s0(u)}
  H0=function(u){-log(s0(u))}
  s=function(b,u) {s0(u)^b}
  h=function(b,u){b*f0(u)/s0(u)}
  H=function(b,u){-b*log(s0(u))}
  root0=function(t){1-pgamma(x0,shape,t)-S0}
```

```
    scale0=uniroot(root0,c(0,10))$root
    scale1=hr
}

G=function(t){1-punif(t, tf, ta+tf)}
g0=function(t){s(scale1,t)*h0(t)*G(t)}
g1=function(t){s(scale1,t)*h(scale1,t)*G(t)}
g00=function(t){s(scale1,t)*H0(t)*h0(t)*G(t)}
g01=function(t){s(scale1,t)*H0(t)*h(scale1,t)*G(t)}
p0=integrate(g0, 0, ta+tf)$value
p1=integrate(g1, 0, ta+tf)$value
p00=integrate(g00, 0, ta+tf)$value
p01=integrate(g01, 0, ta+tf)$value
sigma2.1=p1-p1^2+2*p00-p0^2-2*p01+2*p0*p1
sigma2.0=p0
om=p0-p1

G1=function(t){1-punif(t, t1-ta, t1)}
g0=function(t){s(scale1,t)*h0(t)*G1(t)}
g1=function(t){s(scale1,t)*h(scale1,t)*G1(t)}
g00=function(t){s(scale1,t)*H0(t)*h0(t)*G1(t)}
g01=function(t){s(scale1,t)*H0(t)*h(scale1,t)*G1(t)}
p0=integrate(g0, 0, ta+tf)$value
p1=integrate(g1, 0, ta+tf)$value
p00=integrate(g00, 0, ta+tf)$value
p01=integrate(g01, 0, ta+tf)$value
sigma2.11=p1-p1^2+2*p00-p0^2-2*p01+2*p0*p1
sigma2.01=p0
om1=p0-p1

q1=function(t){s0(t)*h0(t)*G1(t)}
q=function(t){s0(t)*h0(t)*G(t)}
v1=integrate(q1, 0, ta+tf)$value
v=integrate(q, 0, ta+tf)$value
rho0=sqrt(v1/v)
rho1<-sqrt(sigma2.11/sigma2.1)

cL<-(-10)
cU<-(10)
alphac<-100
iter<-0
while ((abs(alphac-alpha)>alphaeps|cU-cL>ceps)&iter<nbmaxiter){
  iter<-iter+1
  c<-(cL+cU)/2
  alphac<-calculate_alpha(c, c1, rho0)
  if (alphac>alpha) {
    cL<-c
  } else {
    cU<-c
  }
}

cb1<-sqrt((sigma2.01/sigma2.11))*(c1-(om1*sqrt(rate*t1)/sqrt(sigma2.01)))
cb<-sqrt((sigma2.0/sigma2.1))*(c-(om*sqrt(rate*ta))/sqrt(sigma2.0))
pwrc<-calculate_power(cb=cb, cb1=cb1, rho1=rho1)
res<-c(cL, cU, alphac, 1-pwrc, rho0, rho1, cb1, cb)
return(res)
```

```
}

  c1<-0 ; rho0<-0; cb1<-0; rho1<-0 ; hz<-c(0,0);
  ceps<-0.001;alphaeps<-0.001;nbmaxiter<-100

Duration=function(shape,S0,x0,hr,tf,rate,alpha,beta,dist)
{
 if (dist=="WB"){
    f0=function(t){(shape/scale0)*(t/scale0)^(shape-1)*exp(-(t/scale0)^shape)}
    s0=function(t){exp(-(t/scale0)^shape)}
    h0=function(t){(shape/scale0)*(t/scale0)^(shape-1)}
    H0=function(t){(t/scale0)^shape}
    s=function(b, t){exp(-b*(t/scale0)^shape)}
    h=function(b,t){b*h0(t)}
    H=function(b,t){b*H0(t)}
    scale0=x0/(-log(S0))^(1/shape)
    scale1=hr
 }


  if (dist=="LN"){
    s0=function(u){1-plnorm(u,scale0,shape)}
    f0=function(u){dlnorm(u,scale0,shape)}
    h0=function(u){f0(u)/s0(u)}
    H0=function(u){-log(s0(u))}
    s=function(b,u) {s0(u)^b}
    h=function(b,u){b*f0(u)/s0(u)}
    H=function(b,u){-b*log(s0(u))}
    scale0=log(x0)-shape*qnorm(1-S0)
    scale1=hr
 }

  if (dist=="LG"){
    s0=function(u){1/(1+(u/scale0)^shape)}
    f0=function(u){(shape/scale0)*(u/scale0)^(shape-1)/(1+(u/scale0)^shape)^2}
    h0=function(u){f0(u)/s0(u)}
    H0=function(u){-log(s0(u))}
    s=function(b,u) {s0(u)^b}
    h=function(b,u){b*f0(u)/s0(u)}
    H=function(b,u){-b*log(s0(u))}
    scale0=x0/(1/S0-1)^(1/shape)
    scale1=hr
 }

  if (dist=="GM"){
    s0=function(u){1-pgamma(u,shape,scale0)}
    f0=function(u){dgamma(u,shape,scale0)}
    h0=function(u){f0(u)/s0(u)}
    H0=function(u){-log(s0(u))}
    s=function(b,u) {s0(u)^b}
    h=function(b,u){b*f0(u)/s0(u)}
    H=function(b,u){-b*log(s0(u))}
    root0=function(t){1-pgamma(x0,shape,t)-S0}
    scale0=uniroot(root0,c(0,10))$root
    scale1=hr
 }
```

```
root=function(ta){
  tau=ta+tf
  G=function(t){1-punif(t, tf, tau)}
  g0=function(t){s(scale1,t)*h0(t)*G(t)}
  g1=function(t){s(scale1,t)*h(scale1,t)*G(t)}
  g00=function(t){s(scale1,t)*H0(t)*h0(t)*G(t)}
  g01=function(t){s(scale1,t)*H0(t)*h(scale1,t)*G(t)}
  p0=integrate(g0, 0, tau)$value
  p1=integrate(g1, 0, tau)$value
  p00=integrate(g00, 0, tau)$value
  p01=integrate(g01, 0, tau)$value
  s1=sqrt(p1-p1^2+2*p00-p0^2-2*p01+2*p0*p1)
  s0=sqrt(p0)
  om=p0-p1
  rate*ta-(s0*qnorm(1-alpha)+s1*qnorm(1-beta))^2/om^2
  }
  ta.single=uniroot(root, lower=0, upper=5*tf)$root
  n.single<-ceiling(ta.single*rate)
  ta.single=ceiling(ta.single)
  ans=list(n.single=n.single, ta.single=ta.single)
  return(ans)
}

 ans=Duration(shape,S0,x0,hr,tf,rate,alpha,beta,dist)
 ta.single<-ans$ta.single
 n.single=ans$n.single

Single_stage<-data.frame(n.single=n.single,ta.single=ta.single,
                         c.single=qnorm(1-alpha))
atc0<-data.frame(n=n.single, t1=ta.single, c1=0.25)
nbpt<-11
pascote<-1.26
cote<-1*pascote
c1.lim<-nbpt*c(-1,1)
EnH0<-10000
iter<-0

while (iter<nbmaxiter & diff(c1.lim)/nbpt>0.001){
iter<-iter+1
cat("iter=",iter,"&EnH0=",round(EnH0, 2),"& Dc1/nbpt=",
    round(diff(c1.lim)/nbpt,5),"\n",sep="")

if (iter%%2==0) nbpt<-nbpt+1
cote<-cote/pascote
n.lim<-atc0$n+c(-1, 1)*n.single*cote
t1.lim<-atc0$t1+c(-1, 1)*ta.single*cote
c1.lim<-atc0$c1+c(-1, 1)*cote
ta.lim<-n.lim/rate
t1.lim<-pmax(0, t1.lim)
n<-seq(n.lim[1], n.lim[2], l=nbpt)
n<-ceiling(n)
n<-unique(n)
ta<-n/rate
t1<-seq(t1.lim[1], t1.lim[2], l=nbpt)
c1<-seq(c1.lim[1], c1.lim[2], l=nbpt)
ta<-ta[ta>0]
t1<-t1[t1>=0.2*ta.single & t1<=1.2*ta.single]
```

```
z<-expand.grid(list(ta=ta, t1=t1, c1=c1))
z<-z[z$ta>z$t1,]
nz<-dim(z)[1]
z$pap<-pnorm(z$c1)
z$eta<-z$ta-pmax(0, z$ta-z$t1)*z$pap
z$enh0<-z$eta*rate
z<-z[z$enh0<=EnH0,]
nz1<-dim(z)[1]
resz<-t(apply(z,1,fct,ceps=ceps,alphaeps=alphaeps,nbmaxiter=nbmaxiter,
         dist=dist))
resz<-as.data.frame(resz)
names(resz)<-c("cL","cU","alphac","betac","rho0","rho1","cb1","cb")
r<-cbind(z, resz)
r$pap<-pnorm(r$c1)
r$eta<-r$ta-pmax(0, r$ta-r$t1)*r$pap
r$etar<-r$eta*rate
r$tar<-r$ta*rate
r$c<-r$cL+(r$cU-resz$cL)/2
r$diffc<-r$cU-r$cL
r$diffc<-ifelse(r$diffc<=ceps, 1, 0)
r<-r[1-r$betac>=1-beta,]
r<-r[order(r$enh0),]
r$n<-r$ta*rate

if (dim(r)[1]>0) {
  atc<-r[, c("ta", "t1", "c1", "n")][1,]

  r1<-r[1,]
  if (r1$enh0<EnH0) {
    EnH0<-r1$enh0
    atc0<-atc
  }
} else {
  atc<-data.frame(ta=NA, t1=NA, c1=NA, n=NA)
}
atc$iter<-iter
atc$enh0<-r$enh0[1]
atc$EnH0<-EnH0
atc$tai<-ta.lim[1]
atc$tas<-ta.lim[2]
atc$ti<-t1.lim[1]
atc$ts<-t1.lim[2]
atc$ci<-c1.lim[1]
atc$cs<-c1.lim[2]
atc$cote<-cote
if (iter==1) {
  atcs<-atc
} else {
  atcs<-rbind(atcs, atc)
}
}
atcs$i<-1:dim(atcs)[1]
a<-atcs[!is.na(atcs$n),]

p<-t(apply(a,1,fct,ceps=ceps,alphaeps=alphaeps,nbmaxiter=nbmaxiter,
      dist=dist))
p<-as.data.frame(p)
```

```
names(p)<-c("cL","cU","alphac","betac","rho0","rho1","cb1","cb")
p$i<-a$i

res<-merge(atcs, p, by="i", all=T)
res$pap<-round(pnorm(res$c1),4)   ## stopping prob of stage 1 ##
res$ta<-res$n/rate
res$EnhO<-(res$ta-pmax(0, res$ta-res$t1)*res$pap)*rate
res$diffc<-res$cU-res$cL
res$c<-round(res$cL+(res$cU-res$cL)/2,4)
res$diffc<-ifelse(res$diffc<=ceps, 1, 0)
res<-res[order(res$enhO),]
des<-round(res[1,],4)

    param<-data.frame(shape=shape,S0=S0,hr=hr,alpha=alpha,beta=beta,
                      rate=rate, x0=x0, tf=tf)
    Two_stage<- data.frame(n1= ceiling(des$t1*param$rate), c1=des$c1,
               n=ceiling(des$ta*param$rate), c=des$c, t1=des$t1,
               MTSL=des$ta+param$tf, ES=des$EnhO, PS=des$pap)
    DESIGN<-list(param=param,Single_stage=Single_stage,Two_stage=Two_stage)
    return(DESIGN)
 }
Optimal.KJ(shape=0.5,S0=0.3,x0=1,hr=0.65,tf=1,rate=10,alpha=0.05,beta=0.2,
           dist="WB")
```

Example 6 *Trial for Advanced Sarcoma*

The data for patients with advanced sarcoma treated with oral cyclophosphamide and sirolimus (OCR) (data are given by Zhao et al., 2011) is used to illustrate two-stage trial design for a new treatment. The PFS survival distribution of OCR trial data (time-to-progression or death is rescaled to month) is estimated by the Kaplan-Meier method and log-logistic distribution which fitted well with a shape parameter $a = 2.3186$ and scale parameter $b = 3.0652$. Thus, we can assume that the survival distribution of the reference group is $S_0(u) = 1/(1 + (u/b)^a)$ and cumulative hazard function is $\Lambda_0(u) = \log(1 + (u/b)^a)$. Assuming that the survival distribution of the experimental group $S_1(u)$ satisfies the following proportional hazards model

$$S_1(u) = [S_0(u)]^\delta,$$

where δ is the hazard ratio. The study aim is to test the following hypothesis:

$$H_0 : S(u) \leq S_0(u) \quad vs. \quad S(u) > S_0(u),$$

with one-sided type I error rate $\alpha = 0.05$ and power of $1 - \beta = 80\%$ to detect an alternative hazard ratio $\delta = 0.65$. The 5 months PFS probability is $S_0(5) = 0.243$ for the reference group. We assume that subjects are uniformly recruited with a constant accrual rate $r = 3$ subjects per month. Suppose followed-up time is $t_f = 6$ months, and no subjects are lost to follow-up, by using R function 'Optimal.KJ', two-stage designs under the log-logistic model is given as follows:

```
$param
    shape    S0   hr alpha beta rate x0 tf
   2.3186 0.243 0.65  0.05  0.2    3  5  6
```

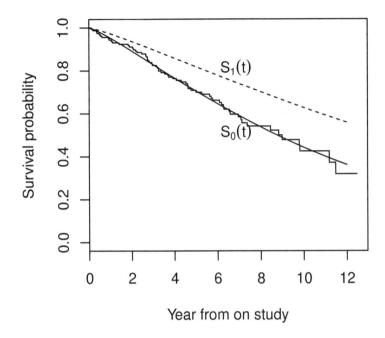

FIGURE 6.2: Kaplan-Meier curve (step function) and fitted log-logistic distribution for OCR trial data.

```
$Single_stage
  n.single ta.single  c.single
        39        13 1.644854

$Two_stage
  n1     c1  n     c     t1    MTSL      ES      PS
  30 -0.2958 43 1.6379 9.9722 20.3333 37.9799 0.3837
```

6.2.4 Simulation

To study characteristics of the proposed optimal two-stage design for the OSLRT, we conducted simulations under various scenarios. In simulations, the survival distribution was taken as the Weibull distribution (WB), log-normal (LN), gamma (GM) and log-logistic (LG) distributions with shape parameter to be 0.5, 1 and 2. The survival probability under null $S_0(x_0)$ at fixed time point $x_0 = 1$ is set to be 0.3; the scale parameters under the null is determined by the value of $S_0(1)$; hazard ratio is set to be $\delta = 0.65$; type I error rate $\alpha = 0.05$ and power of 80%. The simulation results are presented in Table 6.2. The simulation results showed that the two-stage design based OSLRT without restricted follow-up were conservative and slightly overpowered. The study designs were quite robust against the underlying survival distributions.

TABLE 6.1: The characteristics for optimal two-stage design of OSLRT are calculated under Weibull (WB), log-normal (LN), gamma (GM) and log-logistic (LG) distributions by assuming an uniform accrual with accrual rate $r = 10$, the follow-up time $t_f = 1$ or 2 and nominal type I error rate 5% and power of 80%. The empirical type I error and power for the two-stage designs are estimated from 10,000 simulated trials.

Dist.	Shape	$S_0(1)$	δ	t_f	t_1	c_1	c	n_1	n	$\hat{\alpha}$	$1 - \hat{\beta}$
WB	0.5	.3	.65	1	3.10	0.034	1.633	31	49	.040	.838
		.3	.65	2	2.91	-0.064	1.633	30	47	.040	.843
	1	.3	.65	1	2.96	-0.107	1.637	30	44	.037	.840
		.3	.65	2	2.79	-0.219	1.638	28	42	.043	.845
	2	.3	.65	1	2.96	-0.159	1.639	30	42	.042	.843
		.3	.65	2	2.86	-0.232	1.639	29	41	.042	.847
LN	0.5	.3	.65	1	3.01	-0.165	1.639	31	42	.039	.837
		.3	.65	2	2.87	-0.236	1.639	29	41	.040	.842
	1	.3	.65	1	3.00	-0.077	1.636	31	45	.038	.833
		.3	.65	2	2.87	-0.179	1.637	29	43	.039	.838
	2	.3	.65	1	3.12	0.016	1.633	32	50	.041	.833
		.3	.65	2	3.02	-0.017	1.633	31	48	.041	.834
GM	0.5	.3	.65	1	2.97	-0.022	1.633	30	47	.037	.843
		.3	.65	2	2.81	-0.146	1.635	29	44	.037	.845
	1	.3	.65	1	2.96	-0.107	1.637	30	44	.038	.842
		.3	.65	2	2.79	-0.219	1.638	28	42	.038	.840
	2	.3	.65	1	2.95	-0.144	1.638	30	43	.037	.840
		.3	.65	2	2.85	-0.241	1.639	29	41	.038	.842
LG	0.5	.3	.65	1	3.20	0.082	1.632	33	54	.042	.841
		.3	.65	2	3.21	0.057	1.633	33	52	.042	.834
	1	.3	.65	1	3.18	-0.019	1.633	31	49	.040	.834
		.3	.65	2	2.97	-0.061	1.634	30	47	.041	.839
	2	.3	.65	1	2.95	-0.026	1.634	31	45	.038	.835
		.3	.65	2	2.90	-0.159	1.635	29	43	.037	.839

Example 7 *Continuation of Trial for Pancreatic Cancer*

Case and Morgan (2003) illustrated a single-arm phase II trial design to access chemo-radiation treatment for patients with resectable pancreatic cancer. The 12-month survival probability of 35% or less on treatment is considered inefficient, while 12-month survival probability of 50% or greater would be worthwhile. To illustrate two-stage trial design, assume uniform accrual with accrual rate 2 subjects per month and 12 months follow-up; survival times of experimental group follow an exponential distribution $S(t) = e^{-\lambda t}$ with parameter $\lambda = -\log S(12)/12 = -\log(0.5)/12$.

```
Optimal.KJ(shape=1,S0=0.35,hr=log(0.5)/log(0.35),x0=12,rate=2,tf=12,alpha=0.05,
      beta=0.2,dist="WB")
$param
  shape   S0        hr alpha beta rate x0 tf
      1 0.35 0.660252  0.05  0.2    2 12 12
$Single_stage
  nsingle tasingle  csingle
       50       25 1.644854
$Two_stage
  n1     c1  n      c      t1    MTSL      ES      PS
  36 -0.2332 53 1.6281 17.6757  38.5 45.8029 0.4078
```

The two-stage design can be executed as follows: At stage I, we will enroll 36 patients and the interim analysis will be performed after the 36th patient enrolled in the study or at calendar time $t_1 = 17.6757$ (months), whichever occurs later. The stage I test statistic Z_1 will be calculated and compared with stage 1 boundary $c_1 = -0.2332$. If $Z_1 < c_1$, the trial will stop for futility, otherwise, the trial will go to the second stage. At stage II, an additional 17 patients will be enrolled in the study for a total of 53 patients with total study duration $\tau = 38.5$ months. Once the final patient is enrolled in the study and followed for $x = 12$ months, the final analysis will be conducted to calculate the final test statistic Z. If $Z < c_2 = 1.6281$, we will conclude that the new treatment is not worth further investigation. If $Z > c_2$, we will conclude that the new treatment is promising. The operating characteristics of this two-stage design is summarized in Table 6.3. The expected sample size for the two-stage design is 46, the first stage stopping probabilities at null is 0.4078.

To illustrate data analysis after a two-stage designed virtual trial, a virtual sample of 53 survival times of experimental subjects were generated from an exponential distribution $S(t) = e^{-\lambda t}$ with parameter $\lambda = -\log S(12)/12 = -\log(0.5)/12$. We assume patients were uniformly recruited at a rate of 2 patients per month. There was no loss to follow-up, thus, censoring was administrative censoring only. The full follow-up stage I and stage II virtual data are given in Tables 6.4 and 6.5. To calculate the test statistics, we have to calculate the observed number of events and the expected number of events for stage I and stage II data. At stage I, the observed number of events $O_1 = \sum_{i=1}^{n_1} \Delta_i(t_1) = 13$ and the expected number of events $E_1 = \sum_{i=1}^{n_1} \Lambda_0(X_i(t_1)) = 18.741$, where $\Lambda_0(t) = \lambda_0 t$ is the cumulative hazard function under the null with $\lambda_0 = -\log(0.35)/12$. The observed value of

TABLE 6.2: The characteristics for optimal two-stage design of MOSLRT are calculated under Weibull (WB), log-normal (LN), gamma (GM) and log-logistic (LG) distributions by assuming an uniform accrual with accrual rate $r = 10$, the follow-up time $t_f = 1$ or 2 and nominal type I error rate 5% and power of 80%. The empirical type I error and power for the two-stage designs are estimated from 10,000 simulated trials.

Dist.	Shape	$S_0(1)$	δ	t_f	t_1	c_1	c	n_1	n	$\hat{\alpha}$	$1 - \hat{\beta}$
WB	0.5	.3	.65	1	2.90	0.019	1.635	30	44	.046	.826
		.3	.65	2	2.76	-0.164	1.639	28	41	.052	.821
	1	.3	.65	1	2.75	-0.115	1.638	28	40	.046	.826
		.3	.65	2	2.62	-0.331	1.641	27	37	.050	.831
	2	.3	.65	1	2.80	-0.285	1.642	28	37	.050	.830
		.3	.65	2	2.67	-0.399	1.642	27	36	.048	.825
LN	0.5	.3	.65	1	2.80	-0.355	1.643	29	37	.050	.827
		.3	.65	2	2.70	-0.388	1.642	28	36	.049	.822
	1	.3	.65	1	2.89	-0.092	1.639	29	40	.048	.818
		.3	.65	2	2.67	-0.277	1.640	27	38	.049	.820
	2	.3	.65	1	2.92	0.017	1.635	30	45	.056	.828
		.3	.65	2	2.76	-0.101	1.636	28	43	.050	.826
GM	0.5	.3	.65	1	2.84	-0.095	1.639	29	41	.052	.823
		.3	.65	2	2.64	-0.219	1.639	27	39	.048	.829
	1	.3	.65	1	2.75	-0.115	1.638	28	40	.046	.826
		.3	.65	2	2.62	-0.331	1.641	27	37	.050	.831
	2	.3	.65	1	2.76	-0.248	1.642	28	38	.047	.822
		.3	.65	2	2.67	-0.404	1.643	27	36	.048	.823
LG	0.5	.3	.65	1	3.01	0.057	1.635	31	48	.051	.820
		.3	.65	2	3.00	0.021	1.637	30	46	.048	.814
	1	.3	.65	1	2.93	0.011	1.635	30	44	.050	.824
		.3	.65	2	2.80	-0.085	1.637	29	42	.050	.821
	2	.3	.65	1	2.85	-0.095	1.638	29	40	.051	.823
		.3	.65	2	2.67	-0.260	1.639	27	38	.045	.820

TABLE 6.3: The operating characteristics of the two-stage design.

Stage	n	Boundary	Interim time	ES	SP_0
1	36	-0.2332	17.6757		
2	53	1.6281	38.5000	45.8029	0.4078

OSLRT test for stage I is given by

$$Z_1 = \frac{E_1 - O_1}{\sqrt{E_1}} = 1.3262.$$

As $Z_1 > c_1 = -0.2332$, the trial goes to stage II to enroll 17 more patients. After stage II data were collected and full data were analyzed, we have the observed number of events $O = \sum_{i=1}^{n} \Delta_i = 38$ and the expected number of events $E = \sum_{i=1}^{n} \Lambda_0(X_i) = 56.659$. The observed value of OSLRT test for full data is

$$Z = \frac{E - O}{\sqrt{E}} = 2.4789.$$

As $Z > c_2 = 1.6281$, we reject the null hypothesis and conclude that the experimental treatment is promising.

```
#################### Generate Full Follow-up Virtual Data ####################
###   S0 and S1 are survival probabilities at fixed time point x0 under the  ###
###   null and alternative hypothesis; ta and tf are the accrual duration and###
###   follow-up time for the trial. A random sample is generated with size n ###
###   from an exponential distribution under the alternative hypothesis.     ###
##############################################################################
S0=0.35; S1=0.5; x0=12; rate=2; tf=12; n=53
ta=n/rate; lambda=-log(S1)/x0; MTSL=ta+tf
set.seed(83852)
time=rexp(n, lambda)
A=runif(n, 0, ta)
Cens=as.numeric(time<MTSL-A)
Time=pmin(time,MTSL-A)
data=data.frame(Entry=A,Time=Time,Cens=Cens)


############## Data Analysis After Two-Stage Virtual Trial  ##################
###   S0 is survival probabilities at fixed time point x0 under the null;   ###
###   t1 is the stage I interim analysis time; n1 is the sample size for    ###
###   stage I; data is the final data from a two-stage trial. R code 'Test' ###
###   calculates the test statistics z1 at stage I and z at final stage.    ###
##############################################################################
Test=function(S0,x0,t1,tf,n1,data){
 lambda0=-log(S0)/x0
 Lambda0=function(t){lambda0*t}
## order according to entry time ##
 data=data[order(data$Entry),]

## create stage I data ##
 A=data$Entry
 n0=which(A==tail(A[A<t1],1))
 m=max(n0+1,n1)

## interim at n1 or t1 which is later ##
 data2=data[1:m,]
 time1=data2$Time
 A1=data2$Entry
 xt1=pmin(time1,pmax(0,t1-A1))
 delta1=as.numeric(time1<pmax(0,t1-A1))
```

TABLE 6.4: Full follow-up stage I virtual data ($n = 36$) used in the illustrative analysis. Entry : time of entry to the trial (months); Time: time to death or to the analysis (months); Δ: censoring indicator (0=censored, 1=died).

Entry	Time	Δ	Entry	Time	Δ	Entry	Time	Δ	Entry	Time	Δ
0.05	4.28	1	5.16	7.24	1	10.18	1.50	1	15.98	2.16	0
1.13	17.01	0	5.60	2.92	1	10.95	6.68	1	15.99	1.61	1
1.30	16.84	0	5.72	12.42	0	12.04	1.13	1	16.90	1.24	0
1.36	16.78	0	6.35	11.79	0	12.83	5.31	0	17.02	1.12	0
2.14	1.48	1	7.26	10.88	0	14.12	4.02	0	17.02	1.12	0
2.24	15.90	0	7.82	9.71	1	14.47	3.67	0	17.11	1.03	0
3.51	14.63	0	8.59	5.94	1	14.54	3.60	0	17.66	0.48	0
3.52	0.94	1	8.89	9.25	0	14.98	3.16	0	17.85	0.29	0
4.67	12.64	1	9.36	3.16	1	15.84	2.30	0	18.14	0.00	0

```
O1=sum(delta1)
E1=sum(Lambda0(xt1))
Z1=(E1-O1)/sqrt(E1) # stage I test

## final data ##########
Time=data$Time; Cens=data$Cens
O=sum(Cens)
E=sum(Lambda0(Time))
Z=(E-O)/sqrt(E)     # stage II test
ans=c(O1=O1,E1=round(E1,4),Z1=round(Z1,4),
      O=O, E=round(E,4),Z=round(Z,4))
 return(ans)
}
Test(S0=0.35,x=12,tf=12,t1=18.14,n1=36,data=data)
     O1      E1      Z1       O       E       Z
13.0000 18.7415  1.3262 38.0000 56.6593  2.4789
```

6.3 Two-Stage Design with Restricted Follow-up

In previous sections, we have presented the optimal two-stage designs for the trials where patients had full follow-up data. However, in real practice, follow-up information is easier to obtain during the early period of the trial, due to the tight clinical monitoring, as compared to the late period. When the accrual duration is long, the update of the patient's disease progression status at the date of analysis may be challenging because extra site visiting needs to be scheduled for assessing the disease status, while obtaining the disease status of each enrolled patient within a restricted window of time during the

TABLE 6.5: Full follow-up stage II virtual data ($n = 53$) used in the illustrative analysis. Entry : time of entry to the trial (months); T: time to death or to the analysis (months); C: censoring indicator (0=censored, 1=died).

Entry	Time	Δ	Entry	Time	Δ	Entry	Time	Δ	Entry	Time	Δ
0.05	4.28	1	7.82	9.71	1	15.99	1.61	1	22.72	15.78	0
1.13	37.37	0	8.59	5.94	1	16.90	21.60	0	22.73	10.49	1
1.30	23.59	1	8.89	22.89	1	17.02	5.71	1	22.77	0.50	1
1.36	19.64	1	9.36	3.16	1	17.02	4.70	1	23.18	15.32	0
2.14	1.48	1	10.18	1.50	1	17.11	7.41	1	24.49	9.04	1
2.24	18.52	1	10.95	6.68	1	17.66	17.39	1	24.69	11.94	1
3.51	19.21	1	12.04	1.13	1	17.85	11.43	1	25.01	0.80	1
3.52	0.94	1	12.83	5.46	1	18.14	3.65	1	25.06	13.44	0
4.67	12.64	1	14.12	5.00	1	18.58	3.79	1	25.39	13.11	0
5.16	7.24	1	14.47	24.03	0	19.35	4.42	1	25.47	13.03	0
5.60	2.92	1	14.54	23.96	0	19.48	13.76	1	26.34	2.30	1
5.72	32.78	0	14.98	23.52	0	19.61	18.89	0			
6.35	15.42	1	15.84	22.66	0	20.25	11.94	1			
7.26	31.24	0	15.98	22.52	0	21.64	10.18	1			

study period is more realistic. Thus, in this section, we will discuss the optimal two-stage design with restricted follow-up information.

When the data are examined at calendar time $t \leq \tau$, where t is measured from the start of study and τ is the duration of a period of enrollment t_a plus a fixed follow-up time x. Further assuming no loss-to-follow-up during the follow-up period of x, then, we observe the time to failure $X_i(t) = T_i \wedge x \wedge (t - A_i)^+$ and failure indicator $\Delta_i(t) = I(T_i \leq x \wedge (t - A_i)^+), i = 1, \cdots, n$, where A_i is the entry time of the ith subject which is uniformly distributed on interval $[0, t_a]$. Based on the observed data $\{X_i(t), \Delta_i(t), i = 1, \cdots, n\}$, let $N_i(t, u) = \Delta_i(t)I(X_i(t) \leq u)$ and $Y_i(t, u) = I(X_i(t) \geq u)$ be the failure and at-risk process, respectively, where $t \in [0, t_a + x]$ and $u \in [0, x]$. Consider a two-stage design with an interim analysis at calendar time t_1 and final analysis at calendar time $\tau = t_a + x$, corresponding sample sizes are n_1 and n for the interim analysis and final analysis, respectively. At interim analysis t_1, the administrative censoring time is $C_{1i} = (t_1 - A_i)^+$. If the entry time $A_i < t_1$, then $C_{1i} = 0$ and the ith subject has not been enrolled in the study before the interim time t_1; therefore the subject will not be included in the interim analysis. The censoring time $C_{2i} = x$ reflects the amount of follow-up information required by the design. The overall administrative censoring time is $C_i = C_{1i} \wedge C_{2i} = x \wedge (t_1 - A_i)^+$ for the trial with restricted follow-up time x. We assume no loss to follow-up and only administrative censoring for the trial.

6.3.1 Two-Stage Test Statistics

Let $Z_1 = U_1/\hat{\sigma}_1$ and $Z = U/\hat{\sigma}$ be the OSLRT statistics for the first stage and final stage, respectively, where

$$U_1 = n_1^{-1/2} \sum_{i=1}^{n_1} \int_0^x \{Y_i(t_1, u)d\Lambda_0(u) - N_i(t_1, du)\},$$

$$U = n^{-1/2} \sum_{i=1}^{n} \int_0^x \{Y_i(\tau, u)d\Lambda_0(u) - N_i(\tau, du)\}$$

and

$$\hat{\sigma}_1^2 = n_1^{-1} \sum_{i=1}^{n_1} \int_0^x Y_i(t_1, u)d\Lambda_0(u),$$

$$\hat{\sigma}^2 = n^{-1} \sum_{i=1}^{n} \int_0^x Y_i(\tau, u)d\Lambda_0(u),$$

where n_1 and n are the number of subjects enrolled at the stage 1 and final stage. Under the null hypothesis H_0, the means $E(U_1|H_0) = E(U|H_0) = 0$ and variances $\text{var}(U_1|H_0) = \nu_1$ and $\text{var}(U|H_0) = \nu$, where ν_1 and ν are given as follows:

$$\nu_1 = \int_0^x G(u, t_1)S_0(u)d\Lambda_0(u),$$

$$\nu = \int_0^x S_0(u)d\Lambda_0(u),$$

where $G(u, t_1)$ is the survival distribution of censoring time at t_1, that is $G(u, t_1) = P(t_1 - A_1 > u) = P(A_1 < t_1 - u)$ with A_1 is uniformly distributed on $[0, t_a]$. Thus, $G(u, t_1)$ is given as follows:

$$G(u, t_1) = \begin{cases} 1 & u \le t_1 - t_a \\ (t_1 - u)/t_a & t_1 - t_a < u \le t_1 \\ 0 & u > t_1 \end{cases}$$

Under the null hypothesis H_0, by independent increment of the OSLRT, it shows $\text{cov}(U_1, U|H_0) = \text{cov}(U_1, U_1|H_0) = \text{var}(U_1|H_0) = \nu_1$ and $\text{cov}(U, U|H_0) = \text{var}(U|H_0) = \nu$. Furthermore $n_1^{-1} \sum_{i=1}^{n_1} Y_i(t_1, u)$, and $n^{-1} \sum_{i=1}^{n} Y_i(t_2, u)$ uniformly converge to $S_0(u)G(u, t_1)I(u \le x)$ and $S_0(u)I(u \le x)$, respectively, then, $\hat{\sigma}_1^2 \to \nu_1$ and $\hat{\sigma}^2 \to \nu$, or $\text{Var}(Z_1|H_0) = \text{Var}(Z_1|H_0) \simeq 1$. Therefore, the correlation between Z_1 and Z is approximately given as $\rho_0 = \text{cov}(Z_1, Z|H_0) \simeq \text{cov}(U_1, U|H_0)/\sqrt{\nu_1 \nu} = \sqrt{\nu_1/\nu}$. Thus, under the null,$(Z_1, Z)$ is approximately bivariate normal distributed with mean $\mu = (0, 0)'$ and variance matrix

$$\Sigma = \begin{pmatrix} 1 & \rho_0 \\ \rho_0 & 1 \end{pmatrix}$$

Under alternative H_1, we have $E(U_1|H_1) = \sqrt{n_1}\omega_1$ and $E(U|H_1) = \sqrt{n}\omega$, where ω_1 and ω are given as follows:

$$\omega_1 = \int_0^x G(u,t_1)S_1(u)d\Lambda_0(u) - \int_0^x G(u,t_1)S_1(u)d\Lambda_1(u)$$

and

$$\omega = \int_0^x S_1(u)d\Lambda_0(u) - \int_0^x S_1(u)d\Lambda_1(u)$$

and the exact variance of U_1 under H_1 is given by $\sigma_{11}^2 = p_1 - p_1^2 + 2p_{00} - p_0^2 - 2p_{01} + 2p_0 p_1$ with p_0, p_1, p_{00} and p_{01} given as follows:

$$p_0 = \int_0^x G(u,t_1)S_1(u)d\Lambda_0(u),$$

$$p_1 = \int_0^x G(u,t_1)S_1(u)d\Lambda_1(u),$$

$$p_{00} = \int_0^x G(u,t_1)S_1(u)\Lambda_0(u)d\Lambda_0(u),$$

$$p_{01} = \int_0^x G(u,t_1)S_1(u)\Lambda_0(u)d\Lambda_1(u),$$

and the exact variance of U under H_1 is given by $\sigma_{21}^2 = v_1 - v_1^2 + 2v_{00} - v_0^2 - 2v_{01} + 2v_0 v_1$, with v_0, v_1, v_{00} and v_{01} given as follows:

$$v_0 = \int_0^x S_1(u)d\Lambda_0(u),$$

$$v_1 = \int_0^x S_1(x)d\Lambda_1(u),$$

$$v_{00} = \int_0^x S_1(u)\Lambda_0(u)d\Lambda_0(u),$$

$$v_{01} = \int_0^x S_1(u)\Lambda_0(u)d\Lambda_1(u).$$

Similarly, under the alternative hypothesis H_1, by independent increment of the OSLRT, it shows $\text{cov}(U_1, U|H_1) = \text{var}(U_1|H_1) = \sigma_{11}^2$. As the exact variance $\text{var}(U|H_1) = \sigma_{21}^2$, thus, the correlation between U_1 and U under alternative is given by $\rho_1 = \sigma_{11}/\sigma_{21}$. Furthermore, it can be shown $\hat{\sigma}_1^2 \to \sigma_{01}^2 = \int_0^x G(u,t_1)S_1(u)d\Lambda_0(u)$ and $\hat{\sigma}^2 \to \sigma_0^2 = \int_0^x S_1(u)d\Lambda_0(u)$. Therefore, under the alternative, (Z_1, Z) is approximately bivariate normal distributed with mean $\mu = (\sqrt{n_1}\omega_1/\sigma_{01}, \sqrt{n}\omega/\sigma_0)'$ and variance matrix

$$\Sigma = \begin{pmatrix} \sigma_{11}^2/\sigma_{01}^2 & \rho_1\sigma_{11}\sigma_{21}/\sigma_{01}\sigma_0 \\ \rho_1\sigma_{11}\sigma_{21}/\sigma_{01}\sigma_0 & \sigma_{21}^2/\sigma_0^2 \end{pmatrix}$$

For a two-stage design, we consider stop for futility only at the first interim analysis. Suppose that the stopping boundaries are c_1 and c for the first stage and final analysis. At the first stage we stop for futility if $Z_1 \leq c_1$. Otherwise, the trial continues to the second stage and we accept futility or efficacy according to $Z \leq c$ or $Z > c$, respectively, where the boundaries c_1 and c satisfy the following type I error rate α and power $1 - \beta$ constraints:

$$P(Z_1 > c_1, Z > c \,|H_0) = \alpha, \tag{6.6}$$
$$P(Z_1 > c_1, Z > c \,|H_1) = 1 - \beta. \tag{6.7}$$

The two-stage design parameters (n_1, c_1, n, c) are unknown. An iterative algorithm will be implemented to determine (n_1, c_1, n, c) under the following set up of the design parameters.

- Uniform accrual with a constant accrual rate r.

- Survival probability $S_0(x_0)$ at fixed time point x_0.

- Survival distribution function under the null or a shape parameter for a parametric distribution.

- Hazard ratio δ, a proportional hazard model is assumed under the alternative.

- The restricted follow-up time x for each patient.

6.3.2 Optimal Two-Stage Procedure

1. Given $(\alpha, \beta, r, S_0, x_0, \delta, x, \text{dist})$, calculate sample size n_0 and accrual period $t_a^0 = n_0/r$ required for a single-stage design, where r is a constant accrual rate; S_0 is the survival probability under the null at a fixed time point x_0; x is the restricted follow-up time for each patient; "dist" augment is used to specify the underlying survival distribution under the null hypothesis. In our algorithm, we implemented four parametric distributions: Weibull (WB), log-normal (LN), gamma (GM), log-logistic (LG) (see Table 3.1); With a parametric distribution, we only need to specify its shape parameter. Survival distribution under the alternative $S_1(u)$ is determined by proportional hazard assumption, that is $S_1(u) = [S_0(u)]^\delta$, where δ is the hazard ratio.

2. Initial algorithm with $n_1 = n = n_0$, and $c_1 = 0.25$, and given c, calculate α_c by using the following equation (see Appendix D)

$$\alpha_c = \int_c^\infty \phi(z)\Phi\left(\frac{\rho_0 z - c_1}{\sqrt{1 - \rho_0^2}}\right) dz \tag{6.8}$$

and iterate it until α_c is close to the nominal type I error rate α.

3. Calculate the power of a given design (n_1, c_1, n, c) using following equation (see Appendix D for derivation)

$$\text{power} = \int_{\bar{c}}^{\infty} \phi(z)\Phi\left(\frac{\rho_1 z - \bar{c}_1}{\sqrt{1 - \rho_1^2}}\right) dz \qquad (6.9)$$

where

$$\bar{c}_1 = \frac{\sigma_{01}}{\sigma_{11}}\left(c_1 - \frac{\sqrt{n_1}\omega_1}{\sigma_{01}}\right) \quad \text{and} \quad \bar{c} = \frac{\sigma_0}{\sigma_{21}}\left(c - \frac{\sqrt{n}\omega}{\sigma_0}\right)$$

4. If the power is smaller than $1 - \beta$, then (n_1, c_1, n, c) is left and a new triplet (n_1, c_1, n) is tested by repeating steps 2 and 3, otherwise, (n_1, c_1, n, c) is selected.

For each selected candidate design (n_1, c_1, n, c), the expected sample size under H_0 is calculated by $E(n|H_0) = n_1 + (1 - \text{PS}_0)(n - n_1)$, where PS_0 is the probability of early stopping under H_0, that is $\text{PS}_0 = \Phi(c_1)$. Thus, the optimal design is one (n_1, c_1, n, c) which minimizes $E(n|H_0)$. R function 'Optimal.rKJ' implemented above optimal algorithm is given in the next section. 'Optimal.rKJ' refers to the optimal restricted-Kwak and Jung's design.

Remark 1: In the two-stage design, the first stage interim analysis time t_1 is matched to the first stage sample size n_1. However, in real trial, they may not occur at the same time. To insure the power of interim analysis, we suggest that the interim analysis is performed at calendar time t_1 or after n_1 patients were enrolled in the study, whichever occurs later.

Remark 2: Two-stage design with restricted follow-up for the MOSLRT can be developed accordingly.

6.3.3 R Code for Two-Stage Design with Restricted Follow-up

The following R code 'Optimal.rKJ' is modified based on the algorithm developed by Belin et al. (2017). The four parametric survival distributions implemented in the 'Optimal.rKJ' code were the Weibull distribution (WB), log-normal (LN), gamma (GM) and log-logistic (LG) distributions. The R code is implemented under the proportional hazards model, that is $S_1(t) = [S_0(t)]^\delta$, where δ is the hazard ratio.

```
################### Optimal.rKJ Input parameters ############################
###  shape is the shape parameter for one the four parametric distributions;###
###  S0 is the survival probability at fixed time point x0 under the null;  ###
###  hr is inverse of hazard ratio; x is fixed follow-up time period(x>=x0);###
###  rate is constant accrual rate; alpha and beta are type I and II errors ###
###  dist is distribution option with 'WB' as Weibull, 'GM' as Gamma,      ###
```

```
### 'LN' as log-normal, 'LG' as log-logistic.                           ###
##########################################################################
library(survival)
Optimal.rKJ<-function(shape,S0,x0,hr,x,rate,alpha,beta,dist)
{
  calculate_alpha<-function(c2, c1, rho0){
    fun1<-function(z, c1, rho0){
      f<-dnorm(z)*pnorm((rho0*z-c1)/sqrt(1-rho0^2))
      return(f)
    }
    alpha<-integrate(fun1, lower= c2, upper= Inf, c1, rho0)$value
    return(alpha)
  }

  calculate_power<-function(cb, cb1, rho1){
    fun2<-function(z, cb1, rho1){
      f<-dnorm(z)*pnorm((rho1*z-cb1)/sqrt(1-rho1^2))
      return(f)
    }
    pwr<-integrate(fun2,lower=cb,upper=Inf,cb1=cb1,rho1=rho1)$value
    return(pwr)
  }

  fct<-function(zi, ceps=0.0001,alphaeps=0.0001,nbmaxiter=100,dist){
    ta<-as.numeric(zi[1])
    t1<-as.numeric(zi[2])
    c1<-as.numeric(zi[3])

    if (dist=="WB"){
      s0=function(u){1-pweibull(u,shape,scale0)}
      f0=function(u){dweibull(u,shape,scale0)}
      h0=function(u){f0(u)/s0(u)}
      H0=function(u){-log(s0(u))}
      s=function(b,u) {s0(u)^b}
      h=function(b,u){b*f0(u)/s0(u)}
      H=function(b,u){-b*log(s0(u))}
      scale0=x0/(-log(S0))^(1/shape)
      scale1=hr
      ## when b=scale1(=hr) give the survival,
      ## hazard and cumulative hazard at alternative
    }

    if (dist=="LN"){
      s0=function(u){1-plnorm(u,scale0,shape)}
      f0=function(u){dlnorm(u,scale0,shape)}
      h0=function(u){f0(u)/s0(u)}
      H0=function(u){-log(s0(u))}
      s=function(b,u) {s0(u)^b}
      h=function(b,u){b*f0(u)/s0(u)}
      H=function(b,u){-b*log(s0(u))}
      scale0=log(x0)-shape*qnorm(1-S0)
      scale1=hr
    }

    if (dist=="LG"){
      s0=function(u){1/(1+(u/scale0)^shape)}
      f0=function(u){(shape/scale0)*(u/scale0)^(shape-1)/(1+(u/scale0)^shape)^2}
```

```
    h0=function(u){f0(u)/s0(u)}
    H0=function(u){-log(s0(u))}
    s=function(b,u) {s0(u)^b}
    h=function(b,u){b*f0(u)/s0(u)}
    H=function(b,u){-b*log(s0(u))}
    scale0=x0/(1/S0-1)^(1/shape)
    scale1=hr
}

if (dist=="GM"){
    s0=function(u){1-pgamma(u,shape,scale0)}
    f0=function(u){dgamma(u,shape,scale0)}
    h0=function(u){f0(u)/s0(u)}
    H0=function(u){-log(s0(u))}
    s=function(b,u) {s0(u)^b}
    h=function(b,u){b*f0(u)/s0(u)}
    H=function(b,u){-b*log(s0(u))}
    root0=function(t){1-pgamma(x0,shape,t)-S0}
    scale0=uniroot(root0,c(0,10))$root
    scale1=hr
}

g0=function(t){s(scale1,t)*h0(t)}
g1=function(t){s(scale1,t)*h(scale1,t)}
g00=function(t){s(scale1,t)*H0(t)*h0(t)}
g01=function(t){s(scale1,t)*H0(t)*h(scale1,t)}
p0=integrate(g0, 0, x)$value
p1=integrate(g1, 0, x)$value
p00=integrate(g00, 0, x)$value
p01=integrate(g01, 0, x)$value
sigma2.1=p1-p1^2+2*p00-p0^2-2*p01+2*p0*p1
sigma2.0=p0
om=p0-p1

G1=function(t){1-punif(t, t1-ta, t1)}
g0=function(t){s(scale1,t)*h0(t)*G1(t)}
g1=function(t){s(scale1,t)*h(scale1,t)*G1(t)}
g00=function(t){s(scale1,t)*H0(t)*h0(t)*G1(t)}
g01=function(t){s(scale1,t)*H0(t)*h(scale1,t)*G1(t)}
p0=integrate(g0, 0, x)$value
p1=integrate(g1, 0, x)$value
p00=integrate(g00, 0, x)$value
p01=integrate(g01, 0, x)$value
sigma2.11=p1-p1^2+2*p00-p0^2-2*p01+2*p0*p1
sigma2.01=p0
om1=p0-p1

q1=function(t){s0(t)*h0(t)*G1(t)}
q=function(t){s0(t)*h0(t)}
v1=integrate(q1, 0, x)$value
v=integrate(q, 0, x)$value
rho0=sqrt(v1/v)
rho1<-sqrt(sigma2.11/sigma2.1)

cL<-(-10)
cU<-(10)
alphac<-100
```

```
  iter<-0
  while ((abs(alphac-alpha)>alphaeps|cU-cL>ceps)&iter<nbmaxiter){
    iter<-iter+1
    c<-(cL+cU)/2
    alphac<-calculate_alpha(c, c1, rho0)
    if (alphac>alpha) {
      cL<-c
    } else {
      cU<-c
    }
  }

  cb1<-sqrt((sigma2.01/sigma2.11))*(c1-(om1*sqrt(rate*t1)/sqrt(sigma2.01)))
  cb<-sqrt((sigma2.0/sigma2.1))*(c-(om*sqrt(rate*ta))/sqrt(sigma2.0))
  pwrc<-calculate_power(cb=cb, cb1=cb1, rho1=rho1)
  res<-c(cL, cU, alphac, 1-pwrc, rho0, rho1, cb1, cb)
  return(res)
}

  c1<-0 ; rho0<-0; cb1<-0; rho1<-0 ; hz<-c(0,0);
  ceps<-0.001;alphaeps<-0.001;nbmaxiter<-100

  if (dist=="WB"){
    s0=function(u){1-pweibull(u,shape,scale0)}
    f0=function(u){dweibull(u,shape,scale0)}
    h0=function(u){f0(u)/s0(u)}
    H0=function(u){-log(s0(u))}
    s=function(b,u) {s0(u)^b}
    h=function(b,u){b*f0(u)/s0(u)}
    H=function(b,u){-b*log(s0(u))}
    scale0=x0/(-log(S0))^(1/shape)
    scale1=hr
  }

  if (dist=="LN"){
    s0=function(u){1-plnorm(u,scale0,shape)}
    f0=function(u){dlnorm(u,scale0,shape)}
    h0=function(u){f0(u)/s0(u)}
    H0=function(u){-log(s0(u))}
    s=function(b,u) {s0(u)^b}
    h=function(b,u){b*f0(u)/s0(u)}
    H=function(b,u){-b*log(s0(u))}
    scale0=log(x0)-shape*qnorm(1-S0)
    scale1=hr
  }

  if (dist=="LG"){
    s0=function(u){1/(1+(u/scale0)^shape)}
    f0=function(u){(shape/scale0)*(u/scale0)^(shape-1)/(1+(u/scale0)^shape)^2}
    h0=function(u){f0(u)/s0(u)}
    H0=function(u){-log(s0(u))}
    s=function(b,u) {s0(u)^b}
    h=function(b,u){b*f0(u)/s0(u)}
    H=function(b,u){-b*log(s0(u))}
    scale0=x0/(1/S0-1)^(1/shape)
    scale1=hr
  }
```

```
if (dist=="GM"){
  s0=function(u){1-pgamma(u,shape,scale0)}
  f0=function(u){dgamma(u,shape,scale0)}
  h0=function(u){f0(u)/s0(u)}
  H0=function(u){-log(s0(u))}
  s=function(b,u) {s0(u)^b}
  h=function(b,u){b*f0(u)/s0(u)}
  H=function(b,u){-b*log(s0(u))}
  root0=function(t){1-pgamma(x0,shape,t)-S0}
  scale0=uniroot(root0,c(0,10))$root
  scale1=hr
}

g0=function(t){s(scale1,t)*h0(t)}
g1=function(t){s(scale1,t)*h(scale1,t)}
g00=function(t){s(scale1,t)*H0(t)*h0(t)}
g01=function(t){s(scale1,t)*H0(t)*h(scale1,t)}
p0=integrate(g0, 0, x)$value
p1=integrate(g1, 0, x)$value
p00=integrate(g00, 0, x)$value
p01=integrate(g01, 0, x)$value
s1=sqrt(p1-p1^2+2*p00-p0^2-2*p01+2*p0*p1)
s0=sqrt(p0)
om=p0-p1
nsingle<-(s0*qnorm(1-alpha)+s1*qnorm(1-beta))^2/om^2
nsingle<-ceiling(nsingle)
tasingle<-nsingle/rate

Single_stage<-data.frame(n.single=nsingle,ta.single=tasingle,
                         c.single=qnorm(1-alpha))
atc0<-data.frame(n=nsingle, t1=tasingle, c1=0.25)
nbpt<-11
pascote<-1.26
cote<-1*pascote
c1.lim<-nbpt*c(-1,1)
EnH0<-10000
iter<-0

while (iter<nbmaxiter & diff(c1.lim)/nbpt>0.001){
iter<-iter+1
cat("iter=",iter,"&EnH0=",round(EnH0, 2),"& Dc1/nbpt=",
    round(diff(c1.lim)/nbpt,5),"\n",sep="")

if (iter%%2==0) nbpt<-nbpt+1
cote<-cote/pascote
n.lim<-atc0$n+c(-1, 1)*nsingle*cote
t1.lim<-atc0$t1+c(-1, 1)*tasingle*cote
c1.lim<-atc0$c1+c(-1, 1)*cote
ta.lim<-n.lim/rate
t1.lim<-pmax(0, t1.lim)
n<-seq(n.lim[1], n.lim[2], l=nbpt)
n<-ceiling(n)
n<-unique(n)
ta<-n/rate
t1<-seq(t1.lim[1], t1.lim[2], l=nbpt)
c1<-seq(c1.lim[1], c1.lim[2], l=nbpt)
```

```
ta<-ta[ta>0]
t1<-t1[t1>=0.2*tasingle & t1<=1.2*tasingle]
z<-expand.grid(list(ta=ta, t1=t1, c1=c1))
z<-z[z$ta>z$t1,]
nz<-dim(z)[1]
z$pap<-pnorm(z$c1)
z$eta<-z$ta-pmax(0, z$ta-z$t1)*z$pap
z$enh0<-z$eta*rate
z<-z[z$enh0<=EnH0,]
nz1<-dim(z)[1]
resz<-t(apply(z,1,fct,ceps=ceps,alphaeps=alphaeps,nbmaxiter=nbmaxiter,
         dist=dist))
resz<-as.data.frame(resz)
names(resz)<-c("cL","cU","alphac","betac","rho0","rho1","cb1","cb")
r<-cbind(z, resz)
r$pap<-pnorm(r$c1)
r$eta<-r$ta-pmax(0, r$ta-r$t1)*r$pap
r$etar<-r$eta*rate
r$tar<-r$ta*rate
r$c<-r$cL+(r$cU-resz$cL)/2
r$diffc<-r$cU-r$cL
r$diffc<-ifelse(r$diffc<=ceps, 1, 0)
r<-r[1-r$betac>=1-beta,]
r<-r[order(r$enh0),]
r$n<-r$ta*rate

if (dim(r)[1]>0) {
  atc<-r[, c("ta", "t1", "c1", "n")][1,]

  r1<-r[1,]
  if (r1$enh0<EnH0) {
    EnH0<-r1$enh0
    atc0<-atc
  }
} else {
  atc<-data.frame(ta=NA, t1=NA, c1=NA, n=NA)
}
atc$iter<-iter
atc$enh0<-r$enh0[1]
atc$EnH0<-EnH0
atc$tai<-ta.lim[1]
atc$tas<-ta.lim[2]
atc$ti<-t1.lim[1]
atc$ts<-t1.lim[2]
atc$ci<-c1.lim[1]
atc$cs<-c1.lim[2]
atc$cote<-cote
if (iter==1) {
  atcs<-atc
} else {
  atcs<-rbind(atcs, atc)
}
}
atcs$i<-1:dim(atcs)[1]
a<-atcs[!is.na(atcs$n),]

p<-t(apply(a,1,fct,ceps=ceps,alphaeps=alphaeps,nbmaxiter=nbmaxiter,
```

```
       dist=dist))
p<-as.data.frame(p)
names(p)<-c("cL","cU","alphac","betac","rho0","rho1","cb1","cb")
p$i<-a$i

res<-merge(atcs, p, by="i", all=T)
res$pap<-round(pnorm(res$c1),4)  ## stopping prob of stage 1 ##
res$ta<-res$n/rate
res$EnhO<-(res$ta-pmax(0, res$ta-res$t1)*res$pap)*rate
res$diffc<-res$cU-res$cL
res$c<-round(res$cL+(res$cU-res$cL)/2,4)
res$diffc<-ifelse(res$diffc<=ceps, 1, 0)
res<-res[order(res$enh0),]
des<-round(res[1,],4)

if (dist=="SP"){
  param<-data.frame(hr=hr, alpha=alpha, beta=beta, rate=rate, x=x)}
  else {
  param<-data.frame(shape=shape,S0=S0,hr=hr,alpha=alpha,beta=beta,
                 rate=rate, x0=x0, x=x) }
Two_stage<- data.frame(n1= ceiling(des$t1*param$rate), c1=des$c1,
          n=ceiling(des$ta*param$rate), c=des$c, t1=des$t1,
          MTSL=des$ta+param$x, ES=des$EnhO, PS=des$pap)
DESIGN<-list(param=param,Single_stage=Single_stage,Two_stage=Two_stage)
return(DESIGN)
}
Optimal.rKJ(shape=1.47327,S0=0.5,x0=3.5,hr=0.5913,x=5,rate=2,alpha=0.05,
         beta=0.2,dist="WB")
$param
  shape  S0     hr  alpha beta rate  x0 x
1.47327 0.5 0.5914   0.05  0.2    2 3.5 5
$Single_stage
 n.single ta.single  c.single
       42        21  1.644854
$Two_stage
 n1     c1  n      c      t1  MTSL      ES      PS
 28 0.0936 45 1.6269 13.6537  27.5 35.4937 0.5373
```

Example 8 *Continuation of Trial for Pancreatic Cancer with Restricted Follow-up*

Case and Morgan (2003) illustrated a single-arm phase II trial design to access chemo-radiation treatment for patients with resectable pancreatic cancer. The 12-month survival probability of 35% or less on treatment is considered inefficient, while 12-month survival probability of 50% or greater would be worthwhile. To illustrate two-stage trial design, assume uniform accrual with accrual rate 2 subjects per month and with restricted follow-up time 12 months; survival times of experimental group follow an exponential distribution $S(t) = e^{-\lambda t}$ with parameter $\lambda = -\log S(12)/12 = -\log(0.5)/12$.

```
Optimal.rKJ(shape=1,S0=0.35,hr=log(0.5)/log(0.35),x0=12,rate=2,x=12,alpha=0.05,
         beta=0.2,dist="WB")
$param
 shape  S0       hr alpha beta rate x0  x
     1 0.35 0.660252  0.05  0.2    2 12 12
```

TABLE 6.6: The operating characteristics of the two-stage design.

Stage	n	Boundary	Interim time	ES	SP(H_0)
1	44	0.0144	21.9176		
2	73	1.6269	48.5000	58.2514	0.5057

```
$Single_stage
  n.single ta.single  c.single
      69      34.5 1.644854

$Two_stage
  n1    c1  n      c      t1 MTSL      ES      PS
  44 0.0144 73 1.6269 21.9176 48.5 58.2514 0.5057
```

This two-stage design will be executed as follows: At stage I, we will enroll 44 patients and the interim analysis will be performed after the 44th patient is enrolled in the study or at calendar time $t_1 = 21.9176$ (months), whichever occurs later. The stage I test statistic Z_1 will be calculated and compared with stage 1 boundary $c_1 = 0.0144$. If $Z_1 < c_1$, the trial will stop for futility, otherwise, the trial will go to the second stage. At stage II, an additional 29 patients will be enrolled in the study for a total of 73 patients with total study duration MTSL=48.5 months. All patients enrolled in the study are followed for $x = 12$ months (restricted follow-up), the final analysis will be conducted to calculate the final test statistic Z. If $Z < c_2 = 1.6269$, we will conclude that the new treatment is not worthy of further investigation. If $Z > c_2$, we will conclude that the new treatment is promising. The operating characteristics of this two-stage design are summarized in Table 6.6. The expected sample size for the two-stage design is 58.25, the first stage stopping probabilities at null is 0.5057.

To illustrate data analysis after the two-stage virtual trial, a random sample of 73 survival times of experimental subjects were simulated from an exponential distribution $S(t) = e^{-\lambda t}$ with parameter $\lambda = -\log S(12)/12 = -\log(0.5)/12$. Assume patients were uniformly recruited at a rate of 2 patients per month. There was no loss to follow-up, thus, censoring was administrative censoring only. To calculate the test statistics, we have to calculate observed number of events and expected number of events for stage I and stage II data. After 44 patients were enrolled at stage I, the observed number of events is $O_1 = \sum_{i=1}^{n_1} \Delta_i(t_1) = 18$ and the expected number of events is $E_1 = \sum_{i=1}^{n_1} \Lambda_0(X_i(t_1)) = 23.85$, where $\Lambda_0(t) = \lambda_0 t$ is the cumulative hazard function under the null with $\lambda_0 = -\log(0.35)/12$. The observed value of OSLRT test for stage I data is given by

$$Z_1 = \frac{E_1 - O_1}{\sqrt{E_1}} = 1.1979.$$

As $Z_1 > c_1 = 0.0144$, the trial goes to stage II to enroll 29 more patients. After stage II data were collected and analyzed, we have at stage I, the observed number of events $O = \sum_{i=1}^{n} \Delta_i = 40$ and the expected number of events $E = \sum_{i=1}^{n} \Lambda_0(X_i) = 51.53$ The observed value of OSLRT test is

$$Z = \frac{E - O}{\sqrt{E}} = 1.6063.$$

As $Z < c_2 = 1.6281$, we don't reject the null hypothesis and conclude that the experimental treatment is not promising.

```
########### Generate Virtual Survival Data with Restricted Follow-up ##########
###   S0 and S1 are survival probabilities at fixed time point x0 under the  ###
###   null and alternative hypothesis; ta and tf are the accrual duration and###
##    follow-up time for the trial. A random sample with size n from an       ###
###   exponential distribution under the alternative hypothesis is simulated.###
###############################################################################
 S0=0.35; S1=0.5; x0=12; rate=2; x=12; n=73
 ta=n/rate; lambda=-log(S1)/x0; MTSL=ta+x
 set.seed(8385)
 time=rexp(n, lambda)
 A=runif(n, 0, ta)
 Cens=as.numeric(time<pmin(x,MTSL-A))
 Time=pmin(time,pmin(x,MTSL-A))
 data=data.frame(Entry=A,Time=Time,Cens=Cens)


####### Data Analysis Two-Stage Virtual Trial with Restricted Follow-up #######
###   S0 is survival probabilities at fixed time point x0 under the null;    ###
###   t1 is the stage I interim analysis time; n1 is the sample size for     ###
###   stage I; data is the final data from a two-stage trial. R code 'Test'  ###
###   calculates the test statistics z1 at stage I and z at final stage.     ###
###############################################################################
Test=function(S0,x0,x,t1,n1,data){
 lambda0=-log(S0)/x0
 Lambda0=function(t){lambda0*t}
## order according to entry time ##
 data=data[order(data$Entry),]

## create stage I data ##
 A=data$Entry
 n0=which(A==tail(A[A<t1],1))
 m=max(n0+1,n1)

## interim at n1 or t1 which is later ##
 data2=data[1:m,]
 time1=data2$Time
 A1=data2$Entry
 xt1=pmin(time1,pmax(0,pmin(x,t1-A1)))
 delta1=as.numeric(time1<pmax(0,pmin(x,t1-A1)))

 O1=sum(delta1)
 E1=sum(Lambda0(xt1))
 Z1=(E1-O1)/sqrt(E1) # stage I test

## final data ##########
```

```
Time=data$Time; Cens=data$Cens
O=sum(Cens)
E=sum(Lambda0(Time))
Z=(E-O)/sqrt(E)    # stage II test
ans=c(O1=O1,E1=round(E1,4),Z1=round(Z1,4),
      O=O, E=round(E,4),Z=round(Z,4))
 return(ans)
}
Test(S0=0.35,x0=12,x=12,t1=21.9176,n1=44,data=data)
     O1       E1      Z1       O       E       Z
18.0000 23.8500  1.1979 40.0000 51.5305  1.6063
```

Example 9 *Continuation of Trial for Advanced Sarcoma*
The data for patients with advanced sarcoma treated with oral cyclophos-phamide and sirolimus (OCR) (data are given by Zhao et al., 2011) is used to illustrate two-stage trial design for a new treatment. The PFS sur-vival distribution of OCR data (time to progression or death is rescaled to month) is estimated by the Kaplan-Meier method and log-logistic distribu-tion which fitted well with a shape parameter $a = 2.3186$ and scale parameter $b = 3.0652$. Thus, we can assume that the survival distribution of the ref-erence group is $S_0(u) = 1/(1 + (u/b)^a)$ and cumulative hazard function is $\Lambda_0(u) = \log(1 + (u/b)^a)$. Assuming that the survival distribution of the exper-imental group $S_1(u)$ satisfies the following proportional hazards model

$$S_1(u) = [S_0(u)]^\delta,$$

where δ is the hazard ratio. The study aim is to test the following hypothesis:

$$H_0 : S(u) \le S_0(u) \quad vs. \quad S(u) > S_0(u), \quad for\ all\ u \le x$$

with one-sided type I error rate $\alpha = 0.05$ and power of $1 - \beta = 80\%$ to detect an alternative hazard ratio $\delta = 0.65$. The 5 months PFS probability is $S_0(5) = 0.243$ for the reference group. We assume that subjects are uniformly recruited with a constant accrual rate $r = 3$ subjects per month. Suppose each patient is followed $x = 5$ months, and no subjects are lost to follow-up, by using R function 'Optimal.rKJ', two-stage designs under the log-logistic model is given as follows:

```
Optimal.rKJ(shape=2.3186,S0=0.243,x0=5,hr=0.65,x=5,rate=3,alpha=0.05,
            beta=0.2,dist="LG")
$param
   shape     S0   hr alpha beta rate x0 x
  2.3186 0.243 0.65  0.05  0.2    3  5 5

$Single_stage
  n.single ta.single  c.single
        54        18  1.644854

$Two_stage
  n1     c1  n      c      t1    MTSL      ES      PS
  37 0.0585 57 1.6306 12.2521     24 46.4064 0.5233
```

6.3.4   Simulation

To study characteristics of the proposed optimal two-stage design for the OSLRT with restricted follow-up, we conducted simulations under various scenarios. In simulations, the survival distribution was taken as the Weibull distribution (WB), log-normal (LN), gamma (GM) and log-logistic (LG) distributions with shape parameter to be 0.5, 1 and 2. The survival probability under null S_0 at fixed time point $x_0 = 1$ is set to be 0.3; the scale parameters under the null is determined by the value of $S_0(1)$; hazard ratio is set to be $\delta = 0.65$; restricted follow-up time is set to $x = 1$ or 2; type I error rate $\alpha = 0.05$ and power of 80%.

From simulation results presented in Table 6.7, we have the following observations. The proposed optimal two-stage design preserves type I error and power well. The optimal two-stage design increases the maximum sample size from single-stage design by a small amount and the critical value (c) for the final test is also closer to that of the single-stage design ($z_{1-\alpha} = 1.645$). The two-stage designs for different scenarios are similar across the different survival distributions for a short restricted follow-up period ($x = 1$). However, more difference of two-stage designs across four different distributions is expected when the restricted follow-up time becomes longer ($x = 2$).

TABLE 6.7: The characteristics for optimal two-stage design of MOSLRT with restricted follow-up time $x = 1, 2$ were calculated under Weibull (WB), lognormal (LN), gamma (GM) and log-logistic (LG) distributions by assuming an uniform accrual with accrual rate $r = 10$, nominal type I error rate 5% and power of 80%. Overall empirical type I error and power for the two-stage designs were estimated from 10,000 simulated trials.

Dist.	Shape	$S_0(1)$	δ	x	t_1	c_1	c	n_1	n	$\hat{\alpha}$	$1 - \hat{\beta}$
WB	0.5	.3	.65	1	3.44	0.129	1.635	35	55	.054	.814
		.3	.65	2	3.01	0.014	1.638	31	46	.050	.826
	1	.3	.65	1	3.55	0.134	1.634	36	55	.054	.814
		.3	.65	2	2.86	-0.019	1.637	29	41	.049	.823
	2	.3	.65	1	3.52	0.040	1.634	36	55	.054	.814
		.3	.65	2	2.74	-0.279	1.642	28	36	.047	.821
LN	0.5	.3	.65	1	3.56	0.043	1.634	36	55	.050	.812
		.3	.65	2	2.73	-0.292	1.641	28	37	.045	.813
	1	.3	.65	1	3.47	0.061	1.635	35	55	.050	.812
		.3	.65	2	2.85	0.085	1.637	29	42	.050	.818
	2	.3	.65	1	3.43	0.111	1.635	35	55	.050	.812
		.3	.65	2	3.04	0.042	1.637	31	47	.050	.817
GM	0.5	.3	.65	1	3.42	0.096	1.635	35	55	.050	.812
		.3	.65	2	2.90	-0.005	1.637	29	44	.048	.819
	1	.3	.65	1	3.55	0.134	1.634	36	55	.050	.812
		.3	.65	2	2.86	-0.019	1.637	29	41	.051	.818
	2	.3	.65	1	3.51	0.062	1.634	36	55	.050	.812
		.3	.65	2	2.73	-0.222	1.640	28	38	.044	.817
LG	0.5	.3	.65	1	3.44	0.163	1.635	35	55	.050	.812
		.3	.65	2	3.15	0.112	1.635	32	50	.052	.816
	1	.3	.65	1	3.42	0.089	1.635	35	55	.050	.812
		.3	.65	2	3.01	0.020	1.637	31	46	.049	.815
	2	.3	.65	1	3.51	0.068	1.634	36	55	.050	.811
		.3	.65	2	2.85	-0.112	1.638	29	41	.050	.818

7

Phase II Immunotherapy Trial Design

7.1 Introduction

Cancer immunotherapy trials often have two special features. First, because of the indirect mechanism of action of immunotherapy, it takes time for an immune outcome to be elicited and translated into a clinical outcome. Hence, a delayed treatment effect is common to see in survival curves between two groups (e.g., chemo vs. chemo + immuno agent). Second, immune outcome can be sustained after the treatment, thus a proportion of patients may be cured or achieve long-term survival. Both features suggest that the proportional hazards (PH) model may no longer hold true, and using the OSLRT or MOSLRT may lead to loss of efficiency for the study design and data analysis. To be more efficient in designing single-arm phase II immunotherapy trials, we will introduce a weighted MOSLRT and derive a sample size formula.

7.2 Weighted Modified One-Sample Log-Rank Test

Suppose during the accrual phase of the trial, n subjects are enrolled in the study. Let T_i and C_i denote, respectively, the failure time and censoring time of the ith subject. We assume that the failure time T_i and censoring time C_i are independent and $\{T_i, C_i, i = 1, \ldots, n\}$ are independent and identically distributed. Then the observed failure time and failure indicator are $X_i = T_i \wedge C_i$ and $\Delta_i = I(T_i \leq C_i)$, respectively, for the ith subject. Let $N_i(t) = \Delta_i I\{X_i \leq t\}$ and $Y_i(t) = I\{X_i \geq t\}$ be the failure and at-risk processes, respectively. We define

$$O_W = \sum_{i=1}^{n} \int_0^{\infty} W(t)dN_i(t), \quad \text{and} \quad E_W = \sum_{i=1}^{n} \int_0^{\infty} W(t)Y_i(t)d\Lambda_0(t),$$

where $\Lambda_0(t)$ is the cumulative hazard function of the reference group and $W(t)$ is a bounded deterministic weight function. Let $U_W = E_W - O_W$ and then the weighted MOSLRT is defined by

$$L_W = \frac{U_W}{\hat{\sigma}_W}$$

where

$$U_W = n^{-1/2} \sum_{i=1}^{n} \int_0^\infty W(t)\{Y_i(t)d\Lambda_0(t) - dN_i(t)\}$$

and

$$\hat{\sigma}_W^2 = n^{-1} \sum_{i=1}^{n} \int_0^\infty W^2(t)\{Y_i(t)d\Lambda_0(t) + dN_i(t)\}/2.$$

Under the null hypothesis $H_0 : S(t) = S_0(t)$, it can be shown that $\hat{\sigma}_W^2$ convergences in probability to $\int_0^\infty W^2(t)S_0(t)G(t)d\Lambda_0(t)$. By Slutsky's theorem and central limit theorem, the weighted MOSLRT L_W is asymptotically standard normal distributed. Hence, we reject the null hypothesis H_0 with one-sided type I error rate α if $L_W = U_W/\hat{\sigma}_W > z_{1-\alpha}$, where $z_{1-\alpha}$ is the $100(1-\alpha)$ percentile of the standard normal distribution.

Chu et al. (2020) have shown that the asymptotic distribution of U_W under alternative is normal with mean $\sqrt{n}\omega$ and variance σ^2, where

$$\omega = \int_0^\infty W(t)S_1(t)G(t)(d\Lambda_0(t) - d\Lambda_1(t)) \tag{7.1}$$

and

$$\sigma^2 = \sigma_1^2 - \omega^2 - 2\int_0^\infty \left\{\int_0^v W(u)d\Lambda_0(u)\right\} W(v)p(v)(d\Lambda_1(v) - d\Lambda_0(v)) \tag{7.2}$$

with $p(v) = S_1(v)G(v)$ and $\sigma_1^2 = \int_0^\infty W^2(t)S_1(t)G(t)d\Lambda_1(t)$. Further, it can be shown that $\hat{\sigma}_W^2$ converges in probability to $\bar{\sigma}^2 = (\sigma_0^2 + \sigma_1^2)/2$, where

$$\sigma_0^2 = \int_0^\infty W^2(t)S_1(t)G(t)d\Lambda_0(t) \tag{7.3}$$

$$\sigma_1^2 = \int_0^\infty W^2(t)S_1(t)G(t)d\Lambda_1(t). \tag{7.4}$$

Therefore, sample size formula for the weighted MOSLRT is given by

$$n = \frac{(\bar{\sigma}z_{1-\alpha} + \sigma z_{1-\beta})^2}{\omega^2}. \tag{7.5}$$

Remark: When $W(t) = 1$, the sample size formula (7.5) is reduced to the formula (5.9).

7.3 Phase II Trial Design with Delayed Treatment Effect

The proportional hazards model is the most popular model in survival trial design. However, this model is not suitable for immunotherapy cancer trials with delayed treatment effect. In this section, we will introduce a random delayed treatment effect model and develop single-arm phase II trial design using the weighted MOSLRT.

7.3.1 Random Delayed Effect Model

Assume that ζ is a random delay time which is a latent variable and its cumulative distribution function (CDF) is $F_\zeta(t)$ which is assumed to be known. The hazard function of the experimental group can be described by a piecewise proportional hazards model which is given by

$$\lambda_1(t, \zeta) = \begin{cases} \lambda_0(t), & t \leq \zeta \\ \delta\lambda_0(t), & t > \zeta \end{cases}$$

where $\lambda_0(t)$ is the hazard function of the reference group and δ is the hazard ratio of post-delay time ζ. The survival distribution function of the experimental group is given by

$$S_1(t, \zeta) = \begin{cases} S_0(t), & t \leq \zeta \\ [S_0(\zeta)]^{1-\delta}[S_0(t)]^\delta, & t > \zeta \end{cases} \tag{7.6}$$

where $S_0(t)$ is the survival function of the reference group. Since ζ is not observed, we integrate respect to the distribution of ζ to obtain the marginal survival function $S_1(t) = E_\zeta[S_1(t, \zeta)]$ which is given by

$$S_1(t) = S_0(t)S_\zeta(t) + [S_0(t)]^\delta \int_0^t [S_0(u)]^{1-\delta} dF_\zeta(u) \tag{7.7}$$

and the marginal density $f_1(t)$ is given by

$$f_1(t) = f_0(t) \left\{ S_\zeta(t) + \delta[S_0(t)]^{\delta-1} \int_0^t [S_0(u)]^{1-\delta} dF_\zeta(u) \right\}$$

where $f_0(t)$ is the density function of the reference group and $S_\zeta(t) = 1 - F_\zeta(t)$. Figure 7.1 illustrates an example of random delayed treatment effect model with survival curves of the reference group $S_0(t)$ and experimental group $S_1(t)$, and the CDF of a uniform random delay time $F_\zeta(t)$ on interval $[0, 6]$.

7.3.2 Test Statistics

Chu et al. (2020) proposed a $F_\zeta(t)$ weighted OSLRT. The idea is that, at time t with a small value of $F_\zeta(t)$, a small proportion of patients taking immunotherapy has observed treatment effect, and hence, a small weight should

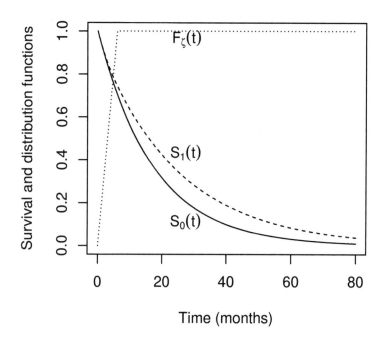

FIGURE 7.1: Survival curves with delayed effect and CDF of a uniform random delay time (dot curve).

be allocated. For a similar reason, at time t with a large value of $F_\zeta(t)$, a large proportion of patients taking immunotherapy has observed treatment effect, and hence a large weight should be allocated. We propose a $F_\zeta(t)$ weighted MOSLRT which is defined as follows:

$$L_{F_\zeta} = \frac{E_{F_\zeta} - O_{F_\zeta}}{\sqrt{(E_{F_\zeta^2} + O_{F_\zeta^2})/2}}$$

where

$$O_{F_\zeta} = \sum_{i=1}^n F_\zeta(X_i)\Delta_i, \quad E_{F_\zeta} = \sum_{i=1}^n \int_0^{X_i} F_\zeta(t)d\Lambda_0(t),$$

and

$$O_{F_\zeta^2} = \sum_{i=1}^n F_\zeta^2(X_i)\Delta_i, \quad E_{F_\zeta^2} = \sum_{i=1}^n \int_0^{X_i} F_\zeta^2(t)d\Lambda_0(t).$$

Let $U_{F_\zeta} = E_{F_\zeta} - O_{F_\zeta}$, using counting-process notations, we have

$$L_{F_\zeta} = \frac{U_{F_\zeta}}{\hat\sigma_{F_\zeta}},$$

where

$$U_{F_\zeta} = n^{-1/2} \sum_{i=1}^n \int_0^\infty F_\zeta(t)\{Y_i(t)d\Lambda_0(t) - dN_i(t)\}$$

and

$$\hat\sigma_{F_\zeta}^2 = n^{-1} \sum_{i=1}^n \int_0^\infty F_\zeta^2(t)\{Y_i(t)d\Lambda_0(t) + dN_i(t)\}/2.$$

Thus, sample size can be calculated by the following formula

$$n = \frac{(\bar\sigma z_{1-\alpha} + \sigma z_{1-\beta})^2}{\omega^2}, \tag{7.8}$$

where $\bar\sigma^2 = (\sigma_0^2 + \sigma_1^2)/2$, where ω, σ^2, σ_0^2 and σ_1^2 can be calculated by equations (7.1), (7.17), (7.3) and (7.4) by replacing the weight function $W(t)$ by $F_\zeta(t)$.

Example 10 *Immunotherapy Trial for Pancreatic Cancer*

We illustrate a single-arm phase II trial design to access immunotherapy for patients with resectable pancreatic cancer. The 12-month survival probability of 35% or less on treatment is considered inefficient, while 12-month survival probability of 50% or greater would be worthwhile. Assume a random delay time of the immunotherapy follows uniform distribution on interval [0, 6] (months), accrual period 34 months and 12 months follow-up; survival times of experimental group follow an exponential distribution $S(t) = e^{-\lambda t}$ with parameter $\lambda = -\log S(12)/12 = -\log(0.5)/12$. A total of $n = 66$ patients are required for the study.

150

150 *Single-Arm Phase II Survival Trial Design*

After the trial for data analysis, we calculate $O_{F_\zeta} = \sum_{i=1}^{n} F_\zeta(X_i)\Delta_i = 65.24$, $E_{F_\zeta} = \sum_{i=1}^{n} \int_0^{X_i} F_\zeta(t)d\Lambda_0(t) = 36.27$, $O_{F_\zeta^2} = \sum_{i=1}^{n} F_\zeta^2(X_i)\Delta_i = 60.74$ and $E_{F_\zeta^2} = \sum_{i=1}^{n} \int_0^{X_i} F_\zeta^2(t)d\Lambda_0(t) = 32.46$. Thus, the test statistic $L = 4.24$ and p-value=0.00001, reject the null hypothesis and conclude that the immunotherapy is promising.

```
Test=function(shape, S0, x0, T1, T2, data)
{
 lambda0=-log(S0)/x0^shape
 h0=function(t){shape*lambda0*t^(shape-1)}
 Ftau=function(t){punif(t, T1, T2)}
 q=function(t){Ftau(t)*h0(t)}
 int.q=function(t){integrate(q, 0, t)$value}
 Time=data$Time
 Cens=data$Cens
 EF=NULL
 for (i in 1:n){
  EF[i]=int.q(Time[i])
 }
 EF=sum(EF)
 q2=function(t){Ftau(t)^2*h0(t)}
 int.q2=function(t){integrate(q2, 0, t)$value}
 EF2=NULL
 for (i in 1:n){
  EF2[i]=int.q2(Time[i])
 }
 EF2=sum(EF2)
 OF=sum(Ftau(Time)*Cens)
 OF2=sum(Ftau(Time)^2*Cens)
 L=(EF-OF)/sqrt((EF2+OF2)/2)
 p=1-pnorm(L)
 return(c(L=round(L,4), p=round(p,6)))
}
Test(shape=1, S0=0.35,x0=12, T1=0, T2=6, data)
        L        p
4.244200 0.000011
```

7.3.3   R Code

```
######## Sample Size Calculation for MOSLRT with Random Delayed Effect ########
### kappa is the Weibull shape parameter; lambda0 is hazard parameter of   ###
### Weibull distribution; alpha and beta are type I and type II errors;    ###
### ta and tf are the accrual and follow-up time: T1 and T2 are the lower  ###
### upper limits of the random delayed time follows uniform dist.          ###
################################################################################
library (statmod)
Size=function(kappa, lambda0, delta, ta, tf, alpha, beta, T1, T2)
{
  quad.points=50
  GQ<-gauss.quad(n=quad.points,  kind="legendre")
  GQ.int<-function(g, limits)
  { upp=limits[2]; low=limits[1];
    sum(sapply(GQ$nodes, function(s)
```

TABLE 7.1: A random delayed virtual data ($n = 66$) were used in the illustrative analysis, where random delay time follows uniform distribution on interval $[0, 6]$ (months). Entry: time of entry to the trial (months); Time: time to death or to data analysis (months); Δ: censoring indicator (0=censored, 1=died).

Entry	Time	Δ	Entry	Time	Δ	Entry	Time	Δ	Entry	Time	Δ
1.35	8.06	1	8.30	1.28	1	17.28	6.70	1	26.78	1.70	1
1.41	44.59	0	8.69	34.21	1	17.56	3.24	1	26.84	19.16	0
2.46	43.54	0	9.67	17.84	1	17.72	8.32	1	27.22	1.35	1
2.88	9.75	1	9.97	17.64	1	18.20	5.43	1	27.24	18.76	0
2.88	27.90	1	11.49	26.94	1	18.26	1.50	1	27.43	18.57	0
3.16	21.89	1	12.48	18.12	1	18.49	27.51	0	27.73	18.27	0
3.50	9.56	1	13.26	0.99	1	19.23	26.77	0	28.23	1.74	1
4.00	4.00	1	13.47	20.22	1	19.45	21.84	1	28.88	17.12	0
6.45	8.66	1	14.15	4.36	1	19.45	0.82	1	28.93	17.07	0
6.56	2.07	1	14.60	2.14	1	20.89	25.11	0	29.97	16.03	0
6.83	11.37	1	14.69	1.00	1	21.21	6.05	1	32.36	0.97	1
7.03	20.12	1	14.76	31.24	0	21.45	16.76	1	32.42	13.58	0
7.24	28.45	1	14.93	8.96	1	23.34	0.84	1	32.54	4.26	1
7.70	5.26	1	15.33	30.67	0	23.55	22.45	0	33.67	0.22	1
7.79	35.76	1	15.41	6.32	1	24.62	18.31	1	33.69	2.83	1
7.91	22.54	1	15.89	13.52	1	24.76	2.29	1			
8.06	1.33	1	17.25	17.65	1	25.34	10.45	1			

```
        {g((upp-low)*s/2+(upp+low)/2)*(upp-low)/2}*GQ$weights)}

z0=qnorm(1-alpha); z1=qnorm(1-beta)
S0=function(t){exp(-lambda0*t^kappa)}
h0=function(t){kappa*lambda0*t^(kappa-1)}
f0=function(t){h0(t)*S0(t)}
G=function(t){1-punif(t, tf, ta+tf)}
F=function(t){punif(t, T1, T2)} ##random lag time uniform on [T1, T2]##
f=function(t){dunif(t, T1, T2)}
int.f=function(t){S0(t)^(1-delta)*f(t)}

S1=function(t){S0(t)*(1-F(t))+S0(t)^delta*GQ.int(int.f, c(0, t))}
f1=function(t){f0(t)*(1-F(t)+delta*S0(t)^(delta-1)*GQ.int(int.f, c(0, t)))}
h1=function(t){f1(t)/S1(t)}
g0=function(t){F(t)^2*S1(t)*G(t)*h0(t)}
g1=function(t){F(t)^2*S1(t)*G(t)*h1(t)}
g10=function(t){F(t)*S1(t)*G(t)*(h1(t)-h0(t))}
sigma0=integrate(g0, 0, ta+tf)$value
omega=integrate(g10, 0, ta+tf)$value
sigma1=integrate(g1, 0, ta+tf)$value
SIGMA1=sigma1-omega^2
int.part=function(v){GQ.int(function(u){F(u)*f0(u)/S0(u)}, c(0,v))}
int.SIGMA=function(v){int.part(v)*F(v)*G(v)*S1(v)*(h1(v)-h0(v))}
bandry=matrix(c(T1,ta+tf), 1,2)
SIGMA2=sum(apply(bandry, 1, function(x) GQ.int(int.SIGMA, limits=x)))
SIGMA=SIGMA1-2*SIGMA2
n=(z0*sqrt((sigma0+sigma1)/2)+z1*sqrt(SIGMA))^2/omega^2
ans=ceiling(n)
return(ans)
}
Size(kappa=1, lambda0=-log(0.35)/12, delta=log(0.5)/log(0.35), ta=34,
    tf=12, alpha=0.05, beta=0.2, T1=0, T2=6)
[1] 66
```

7.4   Phase II Trial Design with Long-Term Survivors

Since immune outcome can be sustained after the treatment, a proportion
of patients may be cured or achieve long-term survival. Therefore, cancer
immunotherapy trials often reflect the mixture of improvement in short-term
risk reduction and long-term survival. In this scenario, a mixture cure model
is useful and hazard functions of the mixture cure model between two groups
may be crossover. Thus, the MOSLRT may be inefficient. To gain efficiency for
the trial design, we propose a change sign weighted MOSLRT for single-arm
phase II trial design with long-term survival.

7.4.1   Mixture Cure Model

The failure time, T^*, is assumed to be $T^* = vT + (1-v)\infty$, where v is an indicator of whether a subject will eventually ($v = 1$) or never ($v = 0$) experience

the failure, and T denotes the failure time if the subject is not cured, with a survival distribution $S(t)$, which is the conditional distribution for patients who will experience the failure, often called the latency distribution. Thus, the marginal survival distribution of T^* is a mixture model of a cure rate $\pi = P(v = 0)$ and a latency distribution $S(t)$ given by

$$S^*(t) = \pi + (1 - \pi)S(t). \tag{7.9}$$

Let $\lambda^*(t)$ and $\Lambda^*(t)$ be the hazard and cumulative hazard functions of T^* and $\lambda(t)$ be the hazard function of T, then we have the following relation for the hazard functions:

$$\lambda^*(t) = \frac{(1 - \pi)S(t)}{\pi + (1 - \pi)S(t)}\lambda(t), \tag{7.10}$$

and for the cumulative hazard function

$$\Lambda^*(t) = -\log\{\pi + (1 - \pi)S(t)\}. \tag{7.11}$$

Let the cure rate and latency distribution be π_0 and $S_0(t)$ for the reference group, π_1 and $S_1(t)$ for the experimental group. We assume a proportional hazards model for the latency survival distribution, that is $S_1(t) = [S_0(t)]^\delta$, where δ is the hazard ratio. Then, the survival distribution, cumulative hazard function and hazard function of the mixture cure rate model are given as follows:

$$
\begin{aligned}
S_i^*(t) &= \pi_i + (1 - \pi_i)S_i(t), \\
\Lambda_i^*(t) &= -\log\{\pi_i + (1 - \pi_i)S_i(t)\}, \\
\lambda_i^*(t) &= \frac{(1 - \pi_i)S_i(t)}{\pi_i + (1 - \pi_i)S_i(t)}\lambda_i(t), \quad i = 0, 1.
\end{aligned}
$$

where $S_1(t) = [S_0(t)]^\delta$ and $\lambda_1(t) = \delta\lambda_0(t)$. A simple mixture cure rate model is an exponential mixture cure rate model

$$S_i^*(t) = \pi_i + (1 - \pi_i)e^{-\lambda_i t}, \tag{7.12}$$

which can be fully specified by the cure rate π_i and hazard rate λ_i for short-term survival.

To detect the treatment effect, we can test the following hypothesis:

$$H_0 : S_1^*(t) = S_0^*(t) \quad vs \quad H_1 : S_1^*(x) > S_0^*(t),$$

or equivalently to testing the following null hypothesis:

$$H_0 : \pi_1 = \pi_0, \quad \text{and} \quad \delta = 1 \tag{7.13}$$

versus various of alternative hypotheses: $H_{1a} : \pi_1 > \pi_0, \delta < 1$, there are differences in both the cure rate and short-term survival; for $H_{1b} : \pi_1 =$

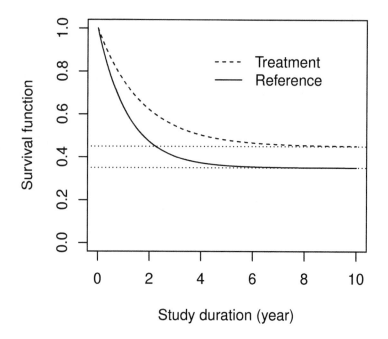

FIGURE 7.2: Survival curves of mixture cure models.

$\pi_0, \delta < 1$, there is a difference in the short-term survival but not in the cure rate; $H_{1c} : \pi_1 > \pi_0, \delta = 1$, there is a difference in the cure rate but not in the short-term survival.

It can be shown if $0 < \pi_0 \leq \pi_1$, and hazard ratio $\delta < 1$, then, there is unique crossing time point of the hazard functions between two groups for the hypotheses H_{1a} and H_{1b} (Appendix E). An example of survival distributions, crossing hazard functions, and non-constant hazard ratio between two treatment groups under mixture cure models for the EOCG trial (see Section 7.4.3) are given in Figures 7.2-7.4. There are no hazards crossing over under the hypothesis H_{1c}. For crossing over cases H_{1a} and H_{1b}, the MOSLRT may be inefficient. Therefore, we propose a change sign weighted MOSLRT for the trial design under the hypotheses H_{1a} and H_{1b} where the hazard functions are crossing over.

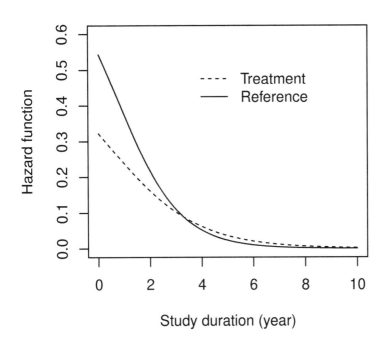

FIGURE 7.3: Crossing over hazard functions of mixture cure models.

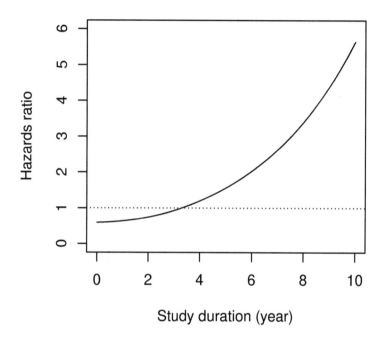

FIGURE 7.4: Non-constant hazard ratio of mixture cure models.

7.4.2 Change Sign Weighted MOSLRT

The U statistics of the MOSLRT can be expressed as follows

$$
\begin{aligned}
U &= n^{-1/2} \sum_{i=1}^{n} \int_0^{\infty} Y_i(t) \left\{ d\Lambda_0^*(t) - \frac{dN_i(t)}{Y_i(t)} \right\} \\
&= n^{-1/2} \sum_{i=1}^{n} \int_0^{\infty} Y_i(t) \{ \lambda_0^*(t) - \hat{\lambda}_1^*(t) \} dt
\end{aligned}
$$

where $\hat{\lambda}_1^*(t)$ is the estimated hazard function under the alternative. Thus, U is a weighted difference of the hazards between the reference and experimental groups. For non-proportionality of crossing hazard functions, the MOSLRT is inefficient for detecting an improvement in long-term survival. Therefore we propose a change sign weighted MOSLRT. Specifically, let t_c be the hazards crossing time point and assuming a positive treatment effect before the crossing point, we then choose following weight function

$$
W(t) = 2\{ I(t \le t_c) - 0.5 \}, \tag{7.14}
$$

which equals $+1$ before the crossing time point and equals -1 after the crossing time point.

To develop test statistics under the mixture cure rate model, suppose during the accrual phase of the trial, n subjects are enrolled in the study. Let T_i^* and C_i denote, respectively, the failure time and censoring time of the ith subject, where the failure time T_i^* follows the mixture cure rate model given in equation (7.9). We assume that the failure time T_i^* and censoring time C_i are independent and $\{T_i^*, C_i, i = 1, \dots, n\}$ are independent and identically distributed. Then the observed failure time and failure indicator are $X_i = T_i^* \wedge C_i$ and $\Delta_i = I(T_i^* \le C_i)$, respectively, for the ith subject. To study the asymptotic distribution of the one-sample test statistic, we formulate it using counting-process notations. Specifically, let $N_i(t) = \Delta_i I\{X_i \le t\}$ and $Y_i(t) = I\{X_i \ge t\}$ be the failure and at-risk processes, respectively. Let

$$
O_W = \sum_{i=1}^{n} \int_0^{\infty} W(t) dN_i(t), \quad E_W = \sum_{i=1}^{n} \int_0^{\infty} W(t) Y_i(t) d\Lambda_0^*(t).
$$

and

$$
O_{W^2} = \sum_{i=1}^{n} \int_0^{\infty} W^2(t) dN_i(t), \quad E_{W^2} = \sum_{i=1}^{n} \int_0^{\infty} W^2(t) Y_i(t) d\Lambda_0^*(t).
$$

where the weight function $W(t)$ is defined by equation (7.14). We define the change sign weighted MOSLRT as follows:

$$
L_W = \frac{E_W - O_W}{\sqrt{(E_{W^2} + O_{W^2})/2}}
$$

then using counting-process notations, we have $L_W = U_W/\hat{\sigma}_W$, where

$$U_W = n^{-1/2} \sum_{i=1}^{n} \int_0^{\infty} W(t)\{Y_i(t)d\Lambda_0^*(t) - dN_i(t)\}$$

and

$$\hat{\sigma}_W^2 = n^{-1} \sum_{i=1}^{n} \int_0^{\infty} W^2(t)\{Y_i(t)d\Lambda_0^*(t) + dN_i(t)\}/2.$$

Similarly, the required sample size for the test statistic L_W is given by

$$n = \frac{(\bar{\sigma}z_{1-\alpha} + \sigma z_{1-\beta})^2}{\omega^2}, \tag{7.15}$$

where

$$\omega = \int_0^{\infty} W(t)S_1^*(t)G(t)(d\Lambda_0^*(t) - d\Lambda_1^*(t)) \tag{7.16}$$

and

$$\sigma^2 = \sigma_1^2 - \omega^2 - 2\int_0^{\infty}\left\{\int_0^{v} W(u)d\Lambda_0^*(u)\right\}W(v)p(v)(d\Lambda_1^*(v) - d\Lambda_0^*(v)) \tag{7.17}$$

with $p(v) = S_1^*(v)G(v)$, $\bar{\sigma}^2 = (\sigma_0^2 + \sigma_1^2)/2$ and

$$\sigma_0^2 = \int_0^{\infty} W^2(t)S_1^*(t)G(t)d\Lambda_0^*(t) \tag{7.18}$$

$$\sigma_1^2 = \int_0^{\infty} W^2(t)S_1^*(t)G(t)d\Lambda_1^*(t). \tag{7.19}$$

Remark: With change sign weight function $W(t)$, we have $W^2(t) = 1$, $O_{W^2} = \sum_{i=1}^{n}\Delta_i$ and $E_{W^2} = \sum_{i=1}^{n}\Lambda_0^*(X_i)$

7.4.3   Study Design and Data Analysis

Example 11 *Trial for Melanoma*
We illustrate the study design under the exponential mixture cure rate model by using the data from the Eastern Cooperative Oncology Group (ECOG) trial e1684 as an example. The ECOG trial e1684 was a two-arm phase III clinical trial to compare the relapse-free survival (RFS) of patients with melanoma who were treated with high-dose interferon alpha-2b (treatment) and placebo as postoperative adjuvant therapy. There were 92 events (relapses) among 146 patients in the treatment group. The SAS macro PSPMCM was applied to this data to fit the treatment arm data under the Weibull cure model, with an estimated shape parameter $\kappa = 1.018$, scale parameter $\lambda_0 = 0.836$ (unit for survival time is year), and cure rate $\pi_0 = 35\%$. Since the

Weibull shape parameter $\kappa = 1.018$ is almost 1, for simplicity, an exponential mixture cure rate model $S_0^*(t) = \pi_0 + (1 - \pi_0)e^{\lambda_0 t}$ is used for the trial design. Suppose an immunotherapy is available, and we want to design a single-arm phase II study to compare the RFS of the immunotherapy to that of the reference group with patients who were treated on the interferon alpha-2b arm of the ECOG trial. The trial is designed to increase the cure rate by 10% ($\pi_1 = 45\%$) and improve the short-term survival by detecting a hazard ratio of 0.7 (Figure 7.5). It can be shown that the hazard functions crossing time point is $t_c = 3.297$. With a one-sided type I error rate $\alpha = 0.05$ and power of 80% at the alternative, 3 years accrual period and 6 years follow-up, using the weighted MOSLRT with weight function $W(t) = 2\{I(t \leq t_c) - 0.5\}$, the required sample size calculated using formula (7.15) under the exponential mixture cure rate model is 60 and the corresponding simulated empirical type I error rate and power are 0.047 and 79.2%, respectively.

To illustrate data analysis after the trial, a random sample of 60 survival times of experimental subjects were simulated from an exponential cure model $S_1(t) = \pi_1 + (1 - \pi_1)e^{-\lambda_1 t}$ with parameter $\lambda_1 = \delta\lambda_0 = 0.5852$ and cure rate $\pi = 0.45$. The data were given in Tables 7.2. To calculate the test statistics, we have to calculate observed number of events and expected number of events for stage I and stage II data. At stage I, the observed number of events $O_W = \sum_{i=1}^{n_1} W(X_i)\Delta_i = 21$ and the expected number of events $E_W = \sum_{i=1}^{n_1} \int_0^{X_i} W(t)\lambda_0^*(t)dt = 39.4816$, $O_{W^2} = \sum_{i=1}^{n} \Delta_i = 33$ and $E_{W^2} = \sum_{i=1}^{n} \Lambda_0^*(X_i) = 45.9991$, where $\lambda_0^*(t) = (1 - \pi_0)S_0(t)/(\pi_0 + (1 - \pi_0)S_0(t))$ and $\Lambda_0^*(t) = -\log(\pi_0 + (1 - \pi_0)S_0(t))$. The observed value of the weighted MOSLRT test is given by

$$Z = \frac{E_W - O_W}{\sqrt{(E_{W^2} + O_{W^2})/2}} = 2.9407$$

and p-value=0.001637, reject the null hypothesis. Thus, we concluded that the new treatment is promising.

```
Test=function(lambda0, pi0, tc, data){
 Time=data$Time; Cens=data$Cens
 A=data$Entry
 S0=function(t){exp(-lambda0*t)}
 h0=function(t){lambda0}
 h0.star=function(t){(1-pi0)*S0(t)*h0(t)/(pi0+(1-pi0)*S0(t))}
 Lambda0.star=function(t){-log(pi0+(1-pi0)*S0(t))}
 W=function(t){2*(1*I(t<=tc)-0.5)}
 O=sum(W(Time)*Cens)
 q=function(t){W(t)*h0.star(t)}
 int.q=function(t){integrate(q, 0, t)$value}
 E=NULL
 for (i in 1:n){
  E[i]=int.q(Time[i])
 }
 E=sum(E)
 O2=sum(Cens)
```

TABLE 7.2: A virtual trial data ($n = 60$) used in the illustrative analysis, where failure time follows a mixture cure rate model. Entry: time of entry to the trial (months); Time: time to death or to the analysis (months); Δ: censoring indicator (0=censored, 1=died).

Entry	Time	Δ	Entry	Time	Δ	Entry	Time	Δ	Entry	Time	Δ
0.01	3.97	1	0.87	8.13	0	1.61	7.39	0	2.23	0.87	1
0.03	2.00	1	0.90	8.10	0	1.62	0.06	1	2.38	6.62	0
0.13	6.30	1	0.93	1.21	1	1.76	7.24	0	2.43	0.52	1
0.15	0.33	1	1.07	0.85	1	1.79	4.77	1	2.49	6.48	1
0.18	8.82	0	1.08	7.92	0	1.82	0.12	1	2.57	6.43	0
0.20	1.50	1	1.15	0.25	1	1.88	7.12	0	2.58	0.69	1
0.28	8.72	0	1.34	7.66	0	1.90	7.10	0	2.61	6.39	0
0.35	2.99	1	1.35	7.65	0	1.92	0.23	1	2.62	3.63	1
0.52	1.05	1	1.36	7.64	0	1.99	7.01	0	2.69	0.19	1
0.60	2.82	1	1.39	1.05	1	2.00	1.60	1	2.78	0.77	1
0.61	8.39	0	1.43	0.73	1	2.05	6.95	0	2.84	0.71	1
0.64	8.36	0	1.51	0.42	1	2.08	6.92	0	2.85	6.15	0
0.73	2.06	1	1.52	7.48	0	2.10	1.28	1	2.95	6.05	0
0.80	3.58	1	1.55	7.45	0	2.18	6.82	0	2.97	6.03	0
0.87	8.13	0	1.56	1.65	1	2.20	1.88	1	2.98	0.99	1

```
E2=sum(Lambda0.star(Time))
Z=(E-O)/sqrt((E2+O2)/2)
ans=c(O=O,E=round(E,4),O2=O2,E2=round(E2,4),Z=round(Z,4))
return(ans)
}
Test(lambda0=0.836, pi0=0.35,tc=3.297, data=data)
       O       E      O2      E2       Z
21.0000 39.4816 33.0000 45.9991  2.9407
```

7.4.4   R Code

```
######## Sample Size Calculation for the Change Sign Weighted MOSLRT ##########
### kappa is the Weibull shape parameter; lambda0 is hazard parameter of    ###
### Weibull distribution; alpha and beta are type I and type II errors;     ###
### ta and tf are the accrual and follow-up time: pi0 and pi1 are the cure  ###
### rates of reference and experimental groups, respectively.               ###
################################################################################
Size=function(kappa, pi0, pi1, lambda0, delta, ta, tf, alpha, power)
{
 S0=function(t){pi0+(1-pi0)*exp(-lambda0*t^kappa)}
 S1=function(t){lambda1=lambda0*delta
  pi1+(1-pi1)*exp(-lambda1*t^kappa)}
 G=function(t){1-punif(t, tf, tau)}
 Lambda0=function(t){-log(S0(t))}
 h0=function(t)
 {lambda0*kappa*t^(kappa-1)*(1-pi0)*exp(-lambda0*t^kappa)/
```

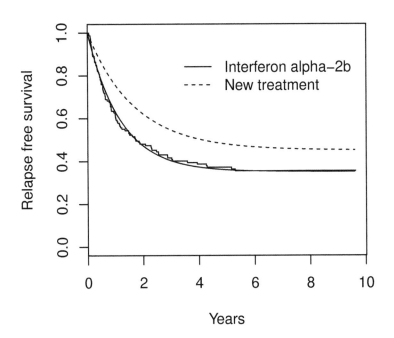

FIGURE 7.5: Kaplan-Meier curve of interferon alpha-2b arm (step function), fitted exponential distribution and hypothetical survival distributions of treatment groups (smooth curves).

```
    (pi0+(1-pi0)*exp(-lambda0*t^kappa))}
h1=function(t){lambda1=lambda0*delta
 lambda1*kappa*t^(kappa-1)*(1-pi1)*exp(-lambda1*t^kappa)/
    (pi1+(1-pi1)*exp(-lambda1*t^kappa))}
f0=function(t){S1(t)*G(t)*h0(t)}
f1=function(t){S1(t)*G(t)*h1(t)}
f2=function(t){G(t)*S1(t)*h0(t)*Lambda0(t)}
f3=function(t){G(t)*S1(t)*h1(t)*Lambda0(t)}
tau=ta+tf
fun=function(t){h1(t)-h0(t)}
tc=uniroot(fun, c(0, 10*tau))$root
w=function(t){2*(1*(t<=tc)-0.5)}
z0=qnorm(1-alpha); z1=qnorm(power)

v0=integrate(f0, 0, tau)$value
v1=integrate(f1, 0, tau)$value
v00=integrate(f2, 0, tau)$value
v01=integrate(f3, 0, tau)$value
g0=function(t){ w(t)*f0(t) }
g1=function(t){ w(t)*f1(t) }
omega=integrate(g0, 0, tau)$value-integrate(g1, 0, tau)$value
n=(sqrt((v0+v1)/2)*z0+sqrt(v1+2*v0*v1-v0^2-v1^2+2*v00-2*v01)*z1)^2/omega^2
ans=ceiling(n); return(ans)
}
Size(kappa=1,lambda0=0.836,pi0=0.35,pi1=0.45,ta=3,tf=6,delta=0.7,
     alpha=0.05,power=0.80)
[1] 60
```

7.4.5   Comparison

We conducted simulation to compare the empirical powers of the MOSLRT (L) and change sign weighted MOSLRT (L_W). Assuming that RFS follows an exponential mixture cure model with median RFS of the latency distribution $m_0 = 12$ (months) for the reference group; patients were accrued to the study uniformly within $t_a = 20$ (months); followed up to $t_f = 10$, 40 and 100 (months) and no loss to follow-up during the study. Sample sizes were calculated with a one-sided type I error rate 5%, power of 80%; cure rate $\pi_0 = 0.4$ for the reference group; cure rate of π_1 from 0.4 to 0.5 for the experimental group; and hazard ratio ranges from 0.5 to 0.7. The empirical type I error rate and power were calculated based on 10,000 simulated trials. From the simulation results we observed the following results. First, both tests preserved type I error rate but slightly liberal for the MOSLRT and the calculated sample sizes provided accurate power for hypothesis H_{1a} (Table 7.3 and Table (7.4). Second, before the hazards crossing time point $t_c = 37$ months (follow-up $t_f = 10$ and study duration $\tau = 30$ months), both tests were efficient and led to nearly identical sample sizes. Thus, if the goal of the study is to detect a short-term risk reduction, the MOSLRT can be used for the trial design and data analysis. However beyond the crossing time point $t_c = 37$ (follow-up $t_f = 40$ and study duration $\tau = 60$ months, or $t_f = 100$ and $\tau = 120$), the MOSLRT lost efficiency while the change sign weighted

MOSLRT gained more efficiency. Thus, if the goal of the study is to detect a long-term survival, we recommend using the change sign weighted MOSLRT for the trial design and data analysis.

For the case of hypothesis H_{1c}, since there is no crossing over, thus change sign weighted MOSLRT (L_W) is the same as the MOSLRT. Therefore, we conducted simulation to study the performance of the MOSLRT. Assuming that RFS follows an exponential mixture cure model with median RFS of the latency distribution $m_0 = 12$ (months) for the reference group; patients were accrued to the study uniformly within $t_a = 20$ (months) and followed up to $t_f = 10$ and 20 (months) and no loss to follow-up during the study. Sample sizes were calculated with a one-sided type I error rate 5%, power of 80%, cure rate $\pi_0 = 0.2$ for the reference group; and cure rate of π_1 from 0.41 to 0.45 for the experimental group; hazard ratio $\delta = 1$. The empirical type I error rate and power were calculated based on 10,000 simulated trials. The simulation results (Table 7.5) showed that the MOSLRT preserved type I error rate and empirical power were all close to the nominal level 80%.

TABLE 7.3: Sample sizes were calculated under the exponential mixture cure model with median 12 (months) and cure rate $\pi_0 = 0.4$ for the reference group, cure rate π_2 ranges from 0.45 to 0.50 for experimental group and hazard ratio ranges from 0.50 to 0.70, accrual duration 20 (months) and follow-up 10, 40 and 100 (months) (case of H_{1a}).

Design							
δ/π_1	t_f	n	$\hat{\alpha}$	EP	n	$\hat{\alpha}$	EP
.70/.45	10	120	.055	79.6	120	.044	79.8
.65/.46	10	83	.055	79.2	83	.043	79.1
.60/.47	10	61	.056	79.5	61	.044	80.0
.55/.48	10	45	.057	79.1	45	.043	79.1
.50/.50	10	33	.060	79.0	33	.045	79.4
.70/.45	40	139	.053	79.8	122	.048	79.7
.65/.46	40	95	.054	79.8	85	.048	79.8
.60/.47	40	67	.054	79.6	61	.047	80.0
.55/.48	40	49	.056	79.4	46	.047	79.9
.50/.50	40	35	.055	80.1	33	.047	79.9
.70/.45	100	222	.053	79.9	97	.048	78.9
.65/.46	100	152	.052	79.5	68	.052	79.0
.60/.47	100	110	.052	79.8	49	.048	78.6
.55/.48	100	81	.054	79.5	37	.044	78.9
.50/.50	100	55	.053	79.5	27	.048	77.9

The column group headers are: Design, L, L_W.

TABLE 7.4: Sample sizes were calculated under exponential mixture cure model with median 12 (months) and cure rate $\pi_0 = \pi_1 = 0.4$ and hazard ratio $\delta = 0.70$, 0.65 and 0.60, accrual duration 20 (months) and follow-up 10, 40 and 100 (months) (case of H_{1b}).

Design			L			L_W	
$\hat{\delta}/\pi_1$	t_f	n	$\hat{\alpha}$	EP	n	$\hat{\alpha}$	EP
.70/.4	10	222	.053	79.8	222	.044	79.4
.65/.4	10	152	.053	79.5	152	.051	79.7
.60/.4	10	108	.054	79.6	108	.046	79.9
.70/.4	40	383	.053	79.8	209	.050	79.1
.65/.4	40	251	.052	79.7	145	.049	79.4
.60/.4	40	170	.053	79.6	105	.047	78.8
.70/.4	100	1130	.050	79.9	134	.052	78.1
.65/.4	100	744	.052	79.7	93	.050	78.0
.60/.4	100	505	.050	79.9	66	.046	77.2

TABLE 7.5: Sample sizes were calculated under exponential mixture cure model with median 12 (months) and cure rate $\pi_0 = 0.2$ for the reference group, cure rate π_2 ranges from 0.41 to 0.45 for experimental group and hazard ratio $\delta = 1$, accrual duration 20 (months) and follow-up 10 and 20 (months) (case of H_{1c}).

Design				L	
δ	π_0/π_1	t_f	n	$\hat{\alpha}$	EP
1	.2/.41	10	75	.054	80.0
1	.2/.42	10	68	.052	79.4
1	.2/.43	10	62	.056	79.8
1	.2/.44	10	57	.053	79.7
1	.2/.45	10	52	.051	80.0
1	.2/.41	20	52	.054	79.8
1	.2/.42	20	47	.054	79.8
1	.2/.43	20	43	.054	79.4
1	.2/.44	20	40	.055	81.0
1	.2/.45	20	36	.056	79.8

7.4.6 Sample Size vs. Length of Follow-up

To explore the relationship between the length of follow-up and sample size for the MOSLRT and change sign weighted MOSLRT, we fixed the study power at 80% and calculated the sample size by varying the length of follow-up from 10 months to 150 months with other design parameters fixed as follows: median survival time 12 months for the reference group, hazard ratio $\delta = 0.7$, cure rate $\pi_0 = 0.4$ and $\pi_1 = 0.5$, uniform accrual with accrual duration 20 months, type I error rate $\alpha = 0.05$. Then, the crossing time point is $t_c = 37$ (months). For the MOSLRT (L), sample size decreased before the hazards crossover and started to increase quickly after the crossing over as the length of follow-up increased. For the change sign weighted MOSLRT (L_W), sample size decreased before the hazards crossing over and started to increase slightly right after the crossing over and then decreased (Figure 7.6).

To explore the relationship between length of follow-up and power, we fix the sample size $n = 120$ and simulated empirical power by varying length of follow-up from 10 months up to 150 months. The empirical power increased slightly before the crossing point for both tests. However after the crossing point, empirical power of the MOSLRT dropped quickly as the length of follow-up increased (drop from 80% to 54%), while the empirical power of the change sign weighted MOSLRT increased as the length of follow-up increased (from 80% to 87%) (Figure 7.7). Thus, through sample size calculation and power simulation, we further confirmed that the MOSLRT lost power with a long-term follow-up after the crossing point. The change sign weighted MOSLRT gained more power with a long-term follow-up after crossing point.

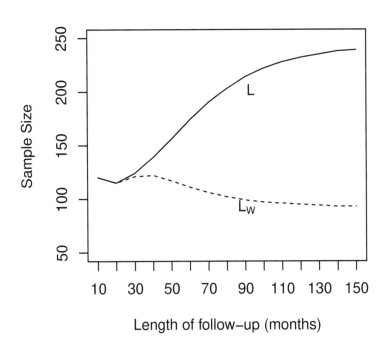

FIGURE 7.6: Sample size vs. length of follow-up.

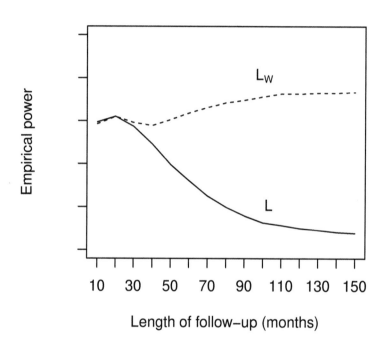

FIGURE 7.7: Empirical power vs. length of follow-up.

8

Phase II Trial Design with GMI Endpoint

8.1 Introduction

For phase II trials of largely these cytostatic therapies where antitumor activity based on tumor shrinkage can't be directly hypothesized, more appropriate endpoints are recommended (Seymour et al., 2010). Specifically, progression-free survival (PFS) or time-to-progression (TTP) have been proposed (Fleming et al., 2009) where PFS is defined as time to disease progression or death. Patients who die without disease progression is censored in the TTP analysis. Von-Hoff (1998) recommended the use of an intra-patient TTP ratio, or growth modulation index (GMI) as the primary endpoint for the trial design. Here the GMI is defined for individual patients as the ratio of their TTP on the current therapy to their TTP on the most recent previous therapy. Thus, patients served as their own control. Because only patients who relapsed or progressed on their most recent therapy will be enrolled in the study, the TTP on the most recent previous therapy, T_1 is uncensored and the TTP on the current therapy, T_2 may be right censored. Therefore the TTP ratio or growth modulation index GMI $= T_2/T_1$ is censored failure time. Furthermore, the TTP pair of (T_1, T_2) is from the same patient, thus, T_1 and T_2 are correlated.

Von-Hoff et al. (2010) proposed to use a simple percentage of GMI > 1.3 as a criterion for drug activity. Using the GMI to make an early assessment of drug efficacy is appealing because it could achieve the dual goals of having a controlled, TTP endpoint based evaluation and a single treatment group design. However, there are several drawbacks for the simple percentage approach. First, it ignores the censoring of T_2, which results in an underestimated probability of GMI > 1.3. Second, the threshold value of 1.3 categorizes the GMI as a binary endpoint and results in an inefficient study design by using the exact binomial test. Third, a study design based on an arbitrary percentage of GMI > 1.3 as a criterion for drug activity is difficult to justify. Even with these drawbacks, this trial endpoint GMI is utilized due to convenience despite the lack of sound statistical methods for designing the trial.

There are few studies in the literature available for the consideration of trial designs with GMI as the primary endpoint. Mick et al. (2000) discussed a hypothesis testing approach for the trial design using a score test and did some simulations under an exponential model. They demonstrated that the correlation between paired TTP times is the key for the efficiency of the trial design. Kovalchik and Mietlowski (2011) proposed an estimation approach. They provided nonparametric and parametric methods to estimate the probability of GMI survival beyond the threshold value. However, their recommendation for the trial design is also based on the simulation results under the Weibull-Gamma frailty model. Similar arguments are discussed by Texier et al. (2018). Recently, Wu et al. (2019b) derived a sample size formula for the score test and discussed study design under a transformed Gumbel's type B bivariate extreme-value (GBVE) model and Weibull frailty model.

8.2   Von-Hoff's Design

Von-Hoff selected threshold GMI > 1.3 for drug activity. To design the study, we define the following probability for evaluating effectiveness of the current therapy

$$p = P(\text{GMI} > 1.3) = S_{\text{GMI}}(1.3),$$

which is the probability that the time-to-progression T_2 of the current therapy is at least 1.3 units of the time-to-progression T_1 of the previous therapy, where $S_{\text{GMI}}(u) = P(\text{GMI} > u)$ is defined as the survival distribution of GMI. Von-Hoff's design is based on testing the following hypothesis

$$H_0 : S_{\text{GMI}}(1.3) \le p_0 \quad vs \quad H_a : S_{\text{GMI}}(1.3) > p_0 \qquad (8.1)$$

and the trial is powered at alternative $S_{\text{GMI}}(1.3) = p_1$. Von-Hoff chose $p_0 = 15\%$ as a futility threshold and $p_1 = 30\%$ as an efficacy threshold. By ignoring censoring of the T_2 (treating censored T_2 observations as uncensored observations) and treating $I(\text{GMI} > 1.3)$ as a binary random variable, the probability $S_{\text{GMI}}(1.3)$ is estimated by a simple proportion of GMI outcomes greater than 1.3, that is

$$\hat{S}_{\text{GMI}}(1.3) = \frac{1}{n} \sum_{i=1}^{n} I(\text{GMI}_i > 1.3),$$

where n is the total sample size (number of patients) and GMI_i is the TTP ratio of ith patient. Its variance is $\sigma^2 = \text{Var}[\hat{S}_{\text{GMI}}(1.3)] = S_{\text{GMI}}(1.3)\{1 - S_{\text{GMI}}(1.3)\}/n$ and variance estimate can be obtained by substituting the point estimate $\hat{S}_{\text{GMI}}(1.3)$. The study design can be conducted using the exact binomial test since the event indicates $\{I(\text{GMI}_i > 1.3), i = 1, \ldots, n\}$ are independent and identically distributed as Bernoulli(p), where $p = S_{\text{GMI}}(1.3)$, e.g., using software STPlan (Brown, et al, 2006). However, ignoring censoring could result in an underestimated survival probability of $p = S_{\text{GMI}}(1.3)$ and inefficient study design.

8.3 Kaplan-Meier Estimate of GMI Distribution

An appropriate approach is to utilize the censored GMI data for the trial design. Assume that the ith subject's TTP for current therapy is T_{2i} and censoring time is C_{2i}, and the observed TTP for the current therapy is $X_{2i} = T_{2i} \wedge C_{2i}$ and failure indicator $\Delta_i = I(T_{2i} \leq C_{2i}), i = 1, \cdots, n$. Let T_{1i} be the TTP of ith subject on previous therapy, which is uncensored, then the observed TTP ratio or GMI for i^{th} subject is $\text{GMI}_i = X_{2i}/T_{1i} = (T_{2i}/T_{1i}) \wedge (C_{2i}/T_{1i})$ and failure indicator is same as $\Delta_i = I(T_{2i} \leq C_{2i})$. Further define failure process $N_i(u) = \Delta_i I(\text{GMI}_i \leq u)$ and at-risk process $Y_i(u) = I(\text{GMI}_i \geq u)$, and $\bar{N}(u) = \sum_{i=1}^{n} N_i(u)$ and $\bar{Y}(u) = \sum_{i=1}^{n} Y_i(u)$. The Kaplan-Meier method can be used to estimate the survival probability $p = S_{\text{GMI}}(1.3)$ as follows

$$\hat{S}_{\text{GMI}}(1.3) = \prod_{u \leq 1.3} \left\{ 1 - \frac{\Delta \bar{N}(u)}{\bar{Y}(u)} \right\},$$

and its variance estimate can be obtained by the Greenwood formula

$$\hat{\text{Var}}[\hat{S}_{\text{GMI}}(1.3)] = \hat{S}_{\text{GMI}}^2(1.3) \int_0^{1.3} \frac{d\bar{N}(u)}{\bar{Y}^2(u)}.$$

To obtain confidence interval of $p = S_{\text{GMI}}(1.3)$, we use log-log transformation which yielding a $(1 - \alpha)\%$ confidence interval

$$\left\{ [\hat{S}_{\text{GMI}}(1.3)]^{exp(\hat{\gamma})}, \ [\hat{S}_{\text{GMI}}(1.3)]^{exp(-\hat{\gamma})} \right\},$$

where

$$\hat{\gamma} = \frac{z_{1-\alpha/2} \sqrt{\hat{\text{Var}}[\hat{S}_{\text{GMI}}(1.3)]}}{|\hat{S}_{\text{GMI}}(1.3) \log \hat{S}_{\text{GMI}}(1.3)|}.$$

Remark: Here time variable of the survival distribution of GMI is no longer the calendar time; instead it represents the value of GMI index. For example 1.3 is the value of GMI index instead of calendar time of the trial.

8.4 Kaplan-Meier Estimate Based Design

To develop sample size calculation for testing the hypothesis (8.1), we can use log-log transformed test statistic

$$Z_1 = \frac{\sqrt{n}\{\log\{-\log(p_0)\} - \log \hat{\Lambda}_{\text{GMI}}(1.3)\} \hat{\Lambda}_{\text{GMI}}(1.3)}{\hat{\sigma}_{\text{GMI}}(1.3)}, \quad (8.2)$$

where $\hat{\Lambda}_{\mathrm{GMI}}(1.3) = -\log\{\hat{S}_{\mathrm{GMI}}(1.3)\}$ and

$$\hat{\sigma}^2_{\mathrm{GMI}}(1.3) = \int_0^{1.3} \frac{d\bar{N}(u)}{\bar{Y}^2(u)/n}.$$

Similar to the sample size calculation in Chapter 5, given type I error rate α and power $1 - \beta$ at the alternative $p_1 = S_{\mathrm{GMI}}(1.3)$, sample size (number of patients) for the test statistic Z_1 can be calculated by the following formula

$$n = \frac{(z_{1-\alpha} + z_{1-\beta})^2 \sigma^2_{\mathrm{GMI}}(1.3)}{[\log\{-\log(p_0)\} - \log\{-\log(p_1)\}]^2 \Lambda^2_{\mathrm{GMI}}(1.3)} \tag{8.3}$$

where $\Lambda_{\mathrm{GMI}}(1.3) = -\log(p_1)$ and asymptotic variance $\sigma^2_{\mathrm{GMI}}(1.3)$ is given by the following equation (8.6).

We can also use the arcsin-square root transformed test statistic

$$Z_2 = \frac{\sqrt{n}\{\arcsin\sqrt{\hat{S}_{\mathrm{GMI}}(1.3)} - \arcsin\sqrt{p_0}\}}{\hat{\nu}}, \tag{8.4}$$

where

$$\hat{\nu}^2 = \frac{\hat{S}_{\mathrm{GMI}}(1.3)\hat{\sigma}^2_{\mathrm{GMI}}(1.3)}{4\{1 - \hat{S}_{\mathrm{GMI}}(1.3)\}}.$$

Sample size (number of patients) can be calculated by the following formula

$$n = \frac{(z_{1-\alpha} + z_{1-\beta})^2 \nu^2(1.3)}{\{\arcsin\sqrt{p_1} - \arcsin\sqrt{p_0}\}^2}, \tag{8.5}$$

where

$$\nu^2(1.3) = \frac{p_1 \sigma^2_{\mathrm{GMI}}(1.3)}{4(1 - p_1)}.$$

To calculate sample size, we have to calculate the asymptotic variance $\sigma^2_{\mathrm{GMI}}(1.3)$ which is given by

$$\sigma^2_{\mathrm{GMI}}(1.3) = \int_0^{1.3} \frac{\lambda_{\mathrm{GMI}}(u)}{\pi(u)} du, \tag{8.6}$$

where $\lambda_{\mathrm{GMI}}(u)$ is the hazard function of GMI, $\pi(u) = P(\mathrm{GMI}_i \geq u) = S_{\mathrm{GMI}}(u)G_{\mathrm{GMI}}(u)$ and $G_{\mathrm{GMI}}(u) = P(C_2/T_1 > u)$ is the censoring survival distribution of GMI. Thus, the asymptotic variance depends on the censoring distribution of GMI which is difficult to specify. Therefore, we assume that there is no censoring to calculate the number of events required for the trial and then by adjusting the censoring rate to calculate the sample size (see Section 8.7). Here an event is defined by $\mathrm{GMI}_i > 1.3$. When there is no censoring, the asymptotic variance $\sigma^2_{\mathrm{GMI}}(1.3)$ can be calculated by

$$\sigma^2_{\mathrm{GMI}}(1.3) = \frac{1}{p_1} - 1.$$

Using this variance calculation, the sample size formulae (8.3) and (8.5) provide the required number of events ($\tilde{n}$) for the current therapy.

Example 12 *Trial for Colon Carcinoma*

The data from a trial reported by Bonetti et al. (2001) is used to illustrate study design using proposed methods. In the trial, treatment efficacy was assessed by GMI for a second-line therapy (combination of LV-modulated 5-FU and oxaliplatin) in patients with advanced colon carcinoma after failure (disease progression) of a first-line chemotherapy (LV-modulated 5-FU or raltitrexed). A scatter plot of 34 observed pairs of TTPs (reconstructed data from Figure 1 of Kovalchik and Mietlowski, 2011) is given in Figure 8.1, which shows a significant correlation between the pairs (correlation $\rho = 0.514$)(the reconstructed data is given in Table 8.1). Sixteen patients (47%) had a GMI > 1.3, twenty-two patients (64.7%) had a GMI > 1 and three patients (9%) had censored T_2 less than T_1. A naive estimate of $p = P(\text{GMI} > 1.3)$ is 47% (by ignoring censoring) while the Kaplan-Meier estimate of $p = P(\text{GMI} > 1.3)$ is 55% (Figure 8.2). If we design the trial by adapting Von-Hoff's method by setting the null hypothesis $p_0 = 15\%$ vs. alternative $p_1 = 30\%$, given type I error rate $\alpha = 0.05$ and power of 80%, the required sample size is 48 by using the exact binomial test while the required sample sizes are 49 and 47 by using the Z_1 and Z_2 test, respectively.

To illustrate data analysis after the trial, we use the data from Bonetti's trial to testing the null hypothesis $H_0 : p = P(\text{GMI} > 1.3) = 15\%$ using the log-log transformed test Z_1 and arcsin square root test Z_2. The Kaplan-Meier estimate is $\hat{p} = 55\%$ (95% CI:39.8% − 75.8%) and p-value calculated from R code "Test" from both tests are approximately 0 which shows strong evidence to reject the null hypothesis.

```
Size=function(p0, p1, alpha, beta)
{
  z0=qnorm(1-alpha); z1=qnorm(1-beta)
  sig2=1/p1-1
  n.Z1=(z0+z1)^2*sig2/((log(-log(p0))-log(-log(p1)))^2*(-log(p1))^2)
  nu=p1*sig2/(4*(1-p1))
  n.Z2=(z0+z1)^2*nu/(asin(sqrt(p0))-asin(sqrt(p1)))^2
  return(c(n.Z1=ceiling(n.Z1),n.Z2=ceiling(n.Z2)))
}
Size(p0=0.15,p1=0.30,alpha=0.05,beta=0.2)
n.Z1 n.Z2
  49   47

Test=function(p0, x0, data)
{
  TTP1=data$TTP1
  TTP2=data$TTP2
  GMI=TTP2/TTP1
  status=data$status
  surv=Surv(GMI, status)
  fit<- survfit(surv ~ 1)
  std.err=tail(summary(fit)$std[summary(fit)$time<=x0],1)
  S.hat=tail(summary(fit)$surv[summary(fit)$time<=x0],1)
  lower=tail(summary(fit)$lower[summary(fit)$time<=x0],1)
  upper=tail(summary(fit)$upper[summary(fit)$time<=x0],1)
```

TABLE 8.1: The data from a trial reported by Bonetti et al. (2001) included 34 observed pairs (T_1, T_2) of time-to-progression for the most recent previous therapy and current therapy and censoring indicator (status) of current therapy.

ID	T_1	T_2	status	ID	T_1	T_2	status
1	4.9	8	1	18	10	31	1
2	10	6.5	1	19	39	52.5	1
3	11.5	6.2	1	20	39	48	1
4	12.8	8	1	21	22.5	45	1
5	13.2	9	1	22	55	47.5	1
6	20	16	1	23	10.8	36.8	1
7	19.5	10	1	24	9.6	34.6	1
8	13.5	14	1	25	8.5	33	1
9	12	17	1	26	9.2	34	1
10	11	21.5	1	27	6	35	1
11	15	25	1	28	17.5	8.5	0
12	20	25	1	29	18.5	15	0
13	22.5	21.5	1	30	12.8	18	0
14	31	21.5	1	31	9.8	20.1	0
15	28	33	1	32	19.5	20	0
16	30	37	1	33	22	46.2	0
17	6.5	31	1	34	31.3	24.2	0

Note: The data are reconstructed from Figure 1 of Kovalchik and Mietlowski, 2011. 0-censoring; 1-progression.

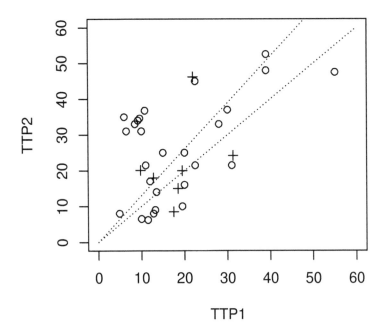

FIGURE 8.1: Scatter plot of T_2 against T_1 for the data from a trial reported by Bonetti et al. (1999) and reconstructed from Figure 1 of Kovalchik and Mietlowski (2011). Dotted lines represent lines of $T_2 = T_1$ and $T_2 = 1.3T_1$ (dot lines). "+" censored observation; "o" uncensored observation.

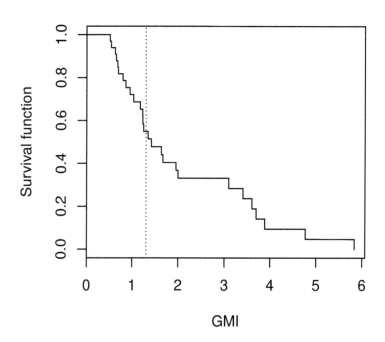

FIGURE 8.2: Kaplan-Meier curve (step function) and line GMI=1.3 (dot line).

```
    nu.hat=std.err^2/(4*S.hat*(1-S.hat))
    Z1=(log(-log(p0))-log(-log(S.hat)))*(-log(S.hat))/(std.err/S.hat)
    Z2=(asin(sqrt(S.hat))-asin(sqrt(p0)))/sqrt(nu.hat)
    p.Z1=round(1-pnorm(Z1), 4)
    p.Z2=round(1-pnorm(Z2), 4)
    return(c(Z1=round(Z1,3), p.Z1=p.Z1, Z2=round(Z2,3), p.Z2=p.Z2,
      S.hat=round(S.hat,4), Lower=round(lower,4), Upper=round(upper,4)))
}
Test(p0=0.15, x0=1.3, data=data)
      Z1    p.Z1      Z2    p.Z2  S.hat   Lower  Upper
  4.2070  0.0000  4.8240  0.0000  0.5495  0.3983  0.7580
```

8.5   Test Quantile of GMI

Both Von-Hoff's design and the Kaplan-Meier estimate based design discussed in Sections 8.2 and 8.3 are based on testing the hypothesis of survival probability $S_{\mathrm{GMI}}(1.3) = p_0$, where the choice of p_0 is arbitrary and difficult to justify its clinical meaning. To overcome this difficulty, we propose an alternative approach of the trial design for testing quantile of the survival distribution of GMI. In general, let q be the θ-quantile of survival function $S_{\mathrm{GMI}}(\cdot)$, that is $q_{\mathrm{GMI}} = S_{\mathrm{GMI}}^{-1}(\theta)$. We consider the following hypothesis

$$H_0 : q_{\mathrm{GMI}} \leq q_0 \quad vs \quad H_1 : q_{\mathrm{GMI}} > q_0. \tag{8.7}$$

When $\theta = 0.5$, it is equivalent to testing the median GMI (m_{GMI}). For the criterion of efficacy, the hypothesis $m_{\mathrm{GMI}} = 1$ vs $m_{\mathrm{GMI}} = 1.3$ means to prolong the median PFS by 30% for the current therapy vs. previous therapy.

The log-log transformed test for testing hypothesis (8.7) is given by

$$Z_1 = \frac{\sqrt{n}\{\log[-\log(\theta)] - \log[\hat{\Lambda}_{\mathrm{GMI}}(q_0)]\}\hat{\Lambda}_{\mathrm{GMI}}(q_0)}{\hat{\sigma}_{\mathrm{GMI}}(q_0)}, \tag{8.8}$$

where $\hat{\Lambda}_{\mathrm{GMI}}(q_0) = -\log\{\hat{S}_{\mathrm{GMI}}(q_0)\}$ and

$$\hat{\sigma}_{\mathrm{GMI}}^2(q_0) = \int_0^{q_0} \frac{d\bar{N}(u)}{\bar{Y}^2(u)/n}.$$

The sample size for the Z_1 test under alternative $H_1 : q_{\mathrm{GMI}} = q_1$ can be calculated by

$$n = \frac{(z_{1-\alpha} + z_{1-\beta})^2 \sigma_{\mathrm{GMI}}^2(q_0)}{\{\log \Lambda_{\mathrm{GMI}}(q_0) - \log[-\log(\theta)]\}^2 \Lambda_{\mathrm{GMI}}^2(q_0)}. \tag{8.9}$$

The arcsin-square root transformation test is given by

$$Z_2 = \frac{\sqrt{n}\{\arcsin\sqrt{\hat{S}_{\mathrm{GMI}}(q_0)} - \arcsin\sqrt{\theta}\}}{\hat{\nu}(q_0)}, \tag{8.10}$$

where

$$\hat{\nu}^2(q_0) = \frac{\hat{S}_{\text{GMI}}(q_0)\hat{\sigma}^2_{\text{GMI}}(q_0)}{4\{1 - \hat{S}_{\text{GMI}}(q_0)\}},$$

and sample size for the Z_2 test can be calculated by

$$n = \frac{(z_{1-\alpha} + z_{1-\beta})^2 \nu^2(q_0)}{\{\arcsin\sqrt{S_{\text{GMI}}(q_0)} - \arcsin\sqrt{\theta}\}^2}, \tag{8.11}$$

where

$$\nu^2(q_0) = \frac{S_{\text{GMI}}(q_0)\sigma^2_{\text{GMI}}(q_0)}{4\{1 - S_{\text{GMI}}(q_0)\}},$$

with $S_{\text{GMI}}(\cdot)$ evaluated at the alternative hypothesis $q_{\text{GMI}} = q_1$. Again, to calculate the sample size, we have to calculate the asymptotic variance $\sigma^2_{\text{GMI}}(q_0)$ which is given by

$$\sigma^2_{\text{GMI}}(q_0) = \int_0^{q_0} \frac{\lambda_{\text{GMI}}(u)}{\pi(u)} du, \tag{8.12}$$

where $\lambda_{\text{GMI}}(u)$ is the hazard function of GMI, $\pi(u) = P(\text{GMI}_i \geq u) = S_{\text{GMI}}(u)G_{\text{GMI}}(u)$. When there is no censoring, the asymptotic variance $\sigma^2_{\text{GMI}}(q_0)$ can be calculated by

$$\sigma^2_{\text{GMI}}(q_0) = \frac{1}{S_{\text{GMI}}(q_0)} - 1,$$

with $S_{\text{GMI}}(\cdot)$ evaluated at the alternative hypothesis $q_{\text{GMI}} = q_1$. Thus, to calculate the sample size, we need to know the survival distribution $S_{\text{GMI}}(\cdot)$ at the alternative hypothesis. In general, the survival distribution $S_{\text{GMI}}(\cdot)$ at the alternative hypothesis is difficult to specify. Therefore, we will discuss two bivariate survival models of (T_1, T_2) to specify the survival distribution $S_{\text{GMI}}(\cdot)$ for the trial design.

8.5.1   Study Design under GBVE Model

We consider a bivariate exponential model which is a transformed Gumbel's type B bivariate extreme-value (GBVE) distribution with a joint bivariate survival distribution function as follows:

$$S(t_1, t_2) = \exp[-\{(t_1/\theta_1)^{1/\nu} + (t_2/\theta_2)^{1/\nu}\}^\nu],$$

where $0 < t_1, t_2 < \infty, 0 < \theta_1, \theta_2 < \infty, 0 < \nu \leq 1$. Here θ_1 and θ_2 are the scale parameters and ν is the dependence parameter, and $\nu = 1$ corresponding to the case of independence. This distribution is referred to as $\text{GBVE}(\theta_1, \theta_2, \nu)$. It is obvious that the marginal distributions of T_1 and T_2 are exponential with scale parameters θ_1 and θ_2, respectively.

Lu and Bhattacharyya (1991) have shown that GMI $= T_2/T_1$ follows a log-logistic distribution of log-logistic(δ, ν), and the survival function of GMI is a log-logistic distribution which is given by

$$S_{\mathrm{GMI}}(s) = \frac{1}{1 + (s\delta)^{1/\nu}}, \tag{8.13}$$

where $\delta = \theta_1/\theta_2$ is the hazard ratio. It can be shown that the Pearson correlation coefficient between T_1 and T_2 is given by

$$\rho = 2\Gamma^2(\nu + 1)/\Gamma(2\nu + 1) - 1,$$

where $\Gamma(\cdot)$ is a gamma-function. For testing θ-quantile of GMI at alternative $q_{\mathrm{GMI}} = q_1$, or $S_{\mathrm{GMI}}(q_1) = \theta$, solving δ from equation (8.13), we have

$$\delta = \frac{1}{q_1} \left(\frac{1}{\theta} - 1 \right)^{\nu}. \tag{8.14}$$

Then, given ν or correlation ρ and θ-quantile q_1 at the alternative, we can obtain δ from equation (8.14) and GMI distribution from equation (8.13). Thus, sample size can be calculated by formula (8.9) or formula (8.11) by using R codes given in Section 8.5.3.

8.5.2 Study Design under Weibull Frailty Model

An alternative model for the trial design with GMI endpoint is the Weibull frailty model (Kovalchik and Mietlowski, 2011). In this approach, we assume that conditionally on a frailty term u, T_1 and T_2 have distributions of Weibull$(u\theta_1, \kappa)$ and Weibull$(u\theta_2, \kappa)$ with a common shape parameter κ, and survival distribution function is given by

$$S(t; \theta_j, \kappa | u) = e^{-(\frac{t}{u\theta_j})^{\kappa}}, \quad j = 1, 2$$

where frailty u has density function $h(u), u > 0$. Owen (2005) has shown that $\log(T_2/T_1)$ follows a logistic distribution of Logistic(γ, κ^{-1}) which does not depend on the shared frailty u, where $\gamma = \log(R)$ and $R = \theta_2/\theta_1$ which is the median ratio. Thus, the survival distribution function of GMI is given by

$$S_{\mathrm{GMI}}(s) = \frac{1}{1 + (s/R)^{\kappa}}. \tag{8.15}$$

For example for testing θ-quantile of GMI is q_1, that is $S_{\mathrm{GMI}}(q_1) = \theta$, solving R we have

$$R = q_1 \left(\frac{1}{\theta} - 1 \right)^{-1/\kappa}. \tag{8.16}$$

Then, given shape parameter κ and θ-quantile q_1 at the alternative, we can obtain R from equation (8.16) and GMI distribution from equation (8.15). Thus, sample size can be calculated by formula (8.9) or formula (8.11) by using R codes given in Section 8.5.3.

TABLE 8.2: Comparison of the Kaplan-Meier estimate based equivalence trial design ($q_0 < 1, q_1 = 1$) using test statistics Z_1 and Z_2 . The total number of events ($\tilde{n}$) are calculated using formula (8.9) with type I error rate $\alpha = 5\%$ and power of 80%. The empirical type I error rate ($\hat{\alpha}$) and empirical power (EP) were calculated based on 10,000 simulation runs under the GBVE model.

Design				Z_1 test			Z_2 test		
ρ	q_0	q_1	θ	$\tilde{n}$	$\hat{\alpha}$	EP	$\tilde{n}$	$\hat{\alpha}$	EP
0.5	0.7	1	0.45	73	.035	82.2	62	.047	79.3
0.6	0.7	1	0.45	55	.035	82.6	45	.060	82.9
0.7	0.7	1	0.45	40	.038	86.7	31	.049	81.4
0.8	0.7	1	0.45	28	.030	88.6	19	.034	75.3
0.5	0.7	1	0.50	76	.049	84.6	63	.062	83.5
0.6	0.7	1	0.50	58	.043	86.6	47	.039	77.8
0.7	0.7	1	0.50	43	.030	85.3	32	.057	81.8
0.8	0.7	1	0.50	30	.052	92.5	20	.055	83.3
0.5	0.7	1	0.55	80	.044	85.6	66	.061	83.5
0.6	0.7	1	0.55	62	.049	88.0	49	.050	81.7
0.7	0.7	1	0.55	46	.031	85.6	34	.046	79.5
0.8	0.7	1	0.55	33	.030	88.4	22	.070	86.6

8.5.3   Comparison

We conducted simulations to study the performance of the proposed two test statistics under the GBVE model. The parameter setting for the simulation is given as follows: correlation ρ is set to be 0.5 to 0.8 by 0.1; θ is set to 0.45, 0.50, and 0.55 and the corresponding θ-quantile is set to $q_0 = 0.7$ under the null $q_1 = 1$ under the alternative; type I error rate $\alpha = 0.05$ and power of 80%. For each design parameter configuration, total number of events $\tilde{n}$ is calculated for each test Z_1 and Z_2, and 10,000 random samples were generated from the GBVE model to estimate empirical type I error rate ($\hat{\alpha}$) and empirical power (EP) which were recorded in Table 8.2 for the case of equivalence trial design ($q_0 = 0.7, q_1 = 1$) and Table 8.3 for the case of efficacy trial design ($q_0 = 1, q_1 = 1.3$). The simulation results showed that the log-log transformed test statistic Z_1 could maintain the type I error rate close to the nominal level but the sample sizes were overestimated. The arcsin-square root transformed test Z_2 provided more accurate sample size estimation but it was also slightly liberal in some cases. Furthermore, the arcsin-square root transformed test Z_2 is more efficient than the log-log transformed Z_1 test.

8.5.4   R Code

```
###### log-log transformed test ##################
Size=function(rho, q0, q1, theta, alpha, beta)
```

TABLE 8.3: Comparison of the Kaplan-Meier estimate based efficacy trial design ($q_0 = 1, q_1 > 1$) using test statistics Z_1 and Z_2 . The total number of events ($\tilde{n}$) are calculated using formula (8.11) with type I error rate $\alpha = 5\%$ and power of 80%. The empirical type I error rate ($\hat{\alpha}$) and empirical power (EP) were calculated based on 10,000 simulation runs under the GBVE model.

	Design			Z_1 test			Z_2 test		
ρ	q_0	q_1	θ	$\tilde{n}$	$\hat{\alpha}$	EP	$\tilde{n}$	$\hat{\alpha}$	EP
0.5	1	1.3	0.45	126	.041	814	113	.051	80.9
0.6	1	1.3	0.45	94	.044	832	82	.048	78.9
0.7	1	1.3	0.45	67	.036	822	56	.046	79.0
0.8	1	1.3	0.45	44	.040	867	34	.036	76.0
0.5	1	1.3	0.50	130	.045	837	115	.047	79.6
0.6	1	1.3	0.50	98	.047	839	84	.049	81.3
0.7	1	1.3	0.50	70	.036	830	58	.043	79.2
0.8	1	1.3	0.50	46	.054	892	35	.044	78.7
0.5	1	1.3	0.55	136	.046	840	118	.059	81.7
0.6	1	1.3	0.55	103	.044	832	87	.050	79.7
0.7	1	1.3	0.55	74	.033	824	60	.046	78.1
0.8	1	1.3	0.55	50	.042	886	37	.043	78.3

```
{ g=function(nu){2*gamma(nu+1)^2/(gamma(2*nu+1))-1-rho}
  nu=uniroot(g, lower=0, upper=1)$root
  S=function(s){1/(1+(s*delta)^(1/nu))}
  delta=(1/q1)*(1/theta-1)^nu
  S1=S(q0); sig2=(1/S1-1)
  z0=qnorm(1-alpha); z1=qnorm(1-beta)
  Lam1=-log(S1)
  n=(z0+z1)^2*sig2/(Lam1^2*(log(Lam1)-log(-log(theta)))^2)
  ans=ceiling(n)
  return(ans)
}
Size(rho=0.5, q0=0.7, q1=1, theta=0.45, alpha=0.05,beta=0.2)
[1] 73
### arcsin square root test ####
Size=function(rho, q0, q1, theta, alpha, beta)
{ g=function(nu){2*gamma(nu+1)^2/(gamma(2*nu+1))-1-rho}
  nu=uniroot(g, lower=0, upper=1)$root
  S=function(s){1/(1+(s*delta)^(1/nu))}
  delta=(1/q1)*(1/theta-1)^nu
  S1=S(q0); sig2=(1/S1-1)
  nu2=S1*sig2/(4*(1-S1))
  z0=qnorm(1-alpha); z1=qnorm(1-beta)
  n=(z0+z1)^2*nu2/(asin(sqrt(S1))-asin(sqrt(theta)))^2
  ans=ceiling(n)
  return(ans)
}
Size(rho=0.5, q0=0.7, q1=1, theta=0.45, alpha=0.05,beta=0.2)
[1] 62
```

8.6   Score Test-Based Design

The treatment activity of the current therapy can be quantified by the hazard ratio of T_2 vs. T_1 as we have seen under the GBVE model and Weibull frailty model. In general, hypothesis testing about the hazard ratio from paired data can be conducted under a log-linear model (Kalbfleisch and Prentice, 2002). In this section, we will use a score test based on a log-linear model for the log TTP ratio and derive a sample size formula for the trial design (Wu et al., 2019).

8.6.1   Score Test Statistics

To assess the treatment effect of the trial with correlated paired survival data $\{(T_{1i}, T_{2i}), i = 1, \ldots, n\}$, a log-linear model for paired difference is considered

$$\log(T_{2i}) - \log(T_{1i}) = \gamma + w_i, \tag{8.17}$$

where $\gamma = -\log(\delta)$ is an acceleration factor for the accelerated failure time (AFT) model (δ is the hazard ratio of T_2 vs. T_1 for the special case of exponential model) and $\{w_i\}$ are independent random errors with a common symmetric distribution. In order to test the hypothesis for the acceleration factor γ, a score is defined for each data pair i as follows:

$$r_i = \begin{cases} 1 & \text{if } T_{2i} > T_{1i} \\ -1 & \text{if } T_{2i} \leq T_{1i} \text{ and } T_{2i} \text{ is uncensored} \end{cases}$$

For the pair i in which $T_{2i} \leq T_{1i}$ and T_{2i} is censored, the pair i does not contribute to the test statistic defined by equation (8.18). Excluding the pairs from the test statistic (8.18) for which r_i can't be specified does not bias the test since r_i takes value -1 and $+1$ with probability 0.5. Let $\tilde{n}$ be the total number of pairs that contribute to the test statistic, we refer to it as the number of paired events. In fact if $T_{2i} > T_{1i}$, whether T_{2i} is uncensored or censored, makes the same contribution to the test statistics. Therefore, it can be treated as an uncensored observation. To test the null hypothesis $H_0 : \gamma = 0$ (or $\delta = 1$), a score test statistic (Kalbfleisch and Prentice, 2002) is defined as follows:

$$Q = \frac{(\sum_{i=1}^{\tilde{n}} r_i)^2}{\sum_{i=1}^{\tilde{n}} r_i^2}, \tag{8.18}$$

which is approximately a central chi-square distribution with degree freedom of 1 (see Appendix Q). Thus, we reject the null hypothesis if $Q > \chi_{1,\alpha}^2$, where $\chi_{1,\alpha}^2$ is the $\alpha\%$ upper quantile of the central chi-square distribution with degree freedom of 1.

8.6.2 Sample Size Formula

To calculate the sample size, we need to derive the asymptotic distribution of the test statistic Q under the alternative $H_a : \delta < 1$. As shown in Appendix F, the test statistic Q follows a non-central chi-square distribution with degree freedom of 1, where the non-central parameter is given as follows:

$$\eta = 4\tilde{n}(p - 0.5)^2,$$

in which $p = P(T_2 > T_1|H_a)$. Thus, given type I error rate α, the study power $1 - \beta$ satisfies

$$1 - \beta = P(Q > \chi^2_{1,\alpha}|H_a) = 1 - F_{1,\eta}(\chi^2_{1,\alpha}),$$

where $F_{1,\eta}(\cdot)$ is the non-central chi-square distribution with degree freedom of 1 and non-central parameter η. Hence, the non-central parameter can be solved from the above equation and defined as $\eta_{\alpha,\beta}$. Therefore, the number of paired events can be calculated by

$$\tilde{n} = \frac{\eta_{\alpha,\beta}}{4(p - 0.5)^2}, \tag{8.19}$$

where $p = P(T_2 > T_1|H_a)$. We discuss the study designs under various models in the next three subsections.

8.6.3 Study Design under GBVE Model

Under the GBVE model, Lu and Bhattacharyya (1991) have shown that $\log(T_2/T_1)$ follows a logistic distribution of $\text{Logistic}(\gamma, \nu)$ with location parameter $\gamma = -\log(\delta)$ ($\delta = \theta_1/\theta_2$) and scale parameter ν, and the survival function of $\log(T_2/T_1)$ is given by

$$S(t; \gamma, \nu) = (1 + e^{\frac{t-\gamma}{\nu}})^{-1}.$$

Thus, the pair (T_1, T_2) satisfies the AFT model (8.17), where the random error w_i follows a symmetric distribution of $\text{Logistic}(0, \nu)$. They have also shown that the Pearson correlation coefficient between T_1 and T_2 is given by

$$\rho = 2\Gamma^2(\nu + 1)/\Gamma(2\nu + 1) - 1,$$

where $\Gamma(\cdot)$ is a gamma-function, and the probability of the growth modulation index GMI > 1 is given by

$$p = P(\text{GMI} > 1) = \frac{1}{1 + \delta^{1/\nu}}, \tag{8.20}$$

where $\delta = \theta_1/\theta_2$ is the hazard ratio. Thus, the study effect size is determined by the correlation ρ and hazard ratio δ. A higher correlation and larger hazard

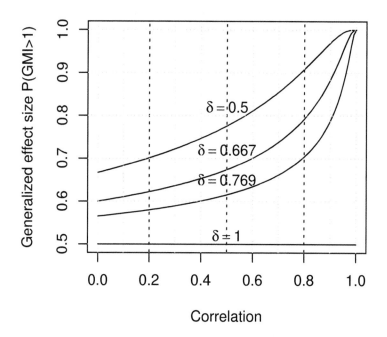

FIGURE 8.3: Graphic representation of the relationship between correlation ρ, hazard ratio δ and generalized effect size $p = P(\text{GMI} > 1)$.

ratio, indicate a larger effect size and higher study power. A graphic presentation of the relationship between the generalized effect size $p = P(\text{GMI} > 1)$ with correlation ρ and hazard ratio δ is given in Figure 8.3.

Given $\rho = 0.3$ to 0.8, and inverse hazard ratio $\delta^{-1} = 1.2$ to 2.5, the required number of events ($\tilde{n}$) calculated for a one-sided type I error of 5% and power of 80% are given in Table 8.4. From sample size calculation, it is again demonstrated that the correlation between paired TTP times is the key for the efficiency of the trial design. Furthermore, sample size calculation also shows that it is unrealistic to design a phase II trial when the correlation is small ($\rho = 0.2 - 0.3$) and hazard ratio is also small ($\delta^{-1} = 1.2 - 1.3$). However, for a moderate correlation ($\rho = 0.4 - 0.6$) and moderate hazard ratio ($\delta^{-1} = 1.4 - 1.6$), less than 100 patients can provide 80% power to detect the significant treatment effect. When hazard ratio is large ($\delta^{-1} \geq 1.7$), then even a small correlation, with less than 100 patients still provides 80% of power for the study design.

TABLE 8.4: The number of paired events ñ required of the trial under the GBVE model for various correlations and hazard ratios for a one-sided hypothesis with type I error of 5% and power of 80%. Based on 100,000 simulation runs, the empirical type I errors ($\hat{\alpha}$) and empirical powers (EP) corresponding to the calculated number of paired events are also recorded in the table.

	Number of paired events ñ ($\hat{\alpha}$/EP)						
	Correlation ρ						
δ^{-1}	0.2	0.3	0.4	0.5	0.6	0.7	0.8
1.2	628(.050/.80)	502(.054/.81)	393(.055/.81)	300(0.056/.82)	219(.042/.79)	150(.041/.78)	92(.047/.81)
1.3	306(.045/.79)	245(.055/.82)	193(.044/.79)	147(.048/.80)	109(.056/.82)	75(.065/.84)	47(.040/.80)
1.4	188(.049/.80)	151(.050/.81)	119(.043/.79)	92(.047/.81)	68(.038/.78)	48(.060/.84)	31(.070/.88)
1.5	131(.053/.82)	106(.041/.79)	84(.063/.84)	65(.046/.81)	49(.043/.81)	35(.041/.81)	23(.035/.82)
1.6	99(.043/.79)	80(.057/.83)	64(.060/.84)	50(.065/.86)	38(.035/.79)	27(.051/.85)	19(.064/.90)
1.7	79(.041/.80)	64(.060/.84)	51(.048/.82)	40(.038/.80)	31(.070/.88)	23(.035/.82)	16(.076/.92)
1.8	65(.046/.81)	53(.053/.83)	43(.067/.86)	34(.057/.85)	26(.075/.89)	19(.064/.88)	14(.057/.91)
1.9	56(.044/.81)	45(.035/.78)	37(.046/.83)	29(.061/.86)	23(.035/.82)	17(.049/.87)	13(.022/.84)
2.0	48(.060/.84)	40(.038/.80)	32(.050/.84)	26(.075/.89)	20(.041/.84)	16(.076/.93)	12(.039/.91)
2.1	43(.067/.86)	35(.041/.80)	29(.061/.87)	23(.035/.81)	18(.031/.81)	14(.057/.90)	11(.064/.95)
2.2	39(.054/.84)	32(.050/.84)	26(.046/.89)	21(.079/.90)	17(.049/.88)	13(.022/.81)	11(.064/.96)
2.3	35(.041/.81)	29(.061/.86)	24(.050/.88)	20(.041/.85)	16(.076/.92)	13(.022/.84)	10(.022/.88)
2.4	32(.050/.83)	27(.051/.85)	22(.061/.86)	18(.031/.81)	15(.035/.86)	12(.039/.90)	10(.022/.91)
2.5	30(.044/.82)	25(.044/.83)	21(.075/.91)	17(.049/.87)	14(.057/.91)	12(.039/.92)	10(.022/.93)

TABLE 8.5: The number of paired events $\tilde{n}$ required of the trial under the Weibull frailty model for one-sided hypothesis with type I error of 5% and power of 80%. Based on 100,000 simulation runs, the empirical type I errors $(\hat{\alpha})$ and empirical powers (EP) corresponding to the calculated number of paired events are also recorded in the table.

		Number of paired events $\tilde{n}$ $(\hat{\alpha}/\text{EP})$				
R	δ^{-1}	$\kappa = 0.5$	δ^{-1}	$\kappa = 1$	δ^{-1}	$\kappa = 2$
1.5	1.2247	769(.051/.80)	1.5	196(.053/.81)	2.25	53(.053/.83)
1.6	1.2649	574(.049/.80)	1.6	147(.048/.80)	2.56	41(.059/.85)
1.7	1.3038	451(.048/.80)	1.7	117(.041/.79)	2.89	33(.035/.79)
1.8	1.3416	369(.048/.80)	1.8	96 (.051/.81)	3.24	28(.036/.80)
1.9	1.3784	310(.048/.80)	1.9	81 (.045/.80)	3.61	24(.064/.87)
2.0	1.4142	267(.051/.81)	2.0	71 (.058/.83)	4.00	22(.051/.87)
2.1	1.4491	233(.049/.80)	2.1	62 (.057/.83)	4.41	20(.042/.85)
2.2	1.4832	207(.053/.81)	2.2	56 (.045/.81)	4.84	18(.030/.82)
2.3	1.5166	186(.047/.80)	2.3	51 (.049/.82)	5.29	17(.049/.88)
2.4	1.5492	169(.046/.79)	2.4	46 (.054/.83)	5.76	16(.077/.92)
2.5	1.5811	155(.054/.82)	2.5	43 (.065/.86)	6.25	15(.036/.86)

8.6.4 Study Design under Weibull Frailty Model

Under the Weibull frailty model, Owen (2005) has shown that $\log(T_2/T_1)$ follows a logistic distribution of $\text{Logistic}(\gamma, \kappa^{-1})$ which does not depend on the shared frailty u, where $\gamma = \log(R)$ and $R = \theta_2/\theta_1$. Thus, the generalized effect size $p = P(T_2 > T_1)$ is given by

$$p = \frac{1}{1 + R^{-\kappa}} = \frac{1}{1 + \delta}, \tag{8.21}$$

where $R = \theta_2/\theta_1$ is the median ratio and $\delta = R^{-\kappa} = (\theta_1/\theta_2)^\kappa$ is the hazard ratio. Given $R = 1.5$ to 2.5 and $\kappa = 0.5$, 1 and 2 to reflect cases of decreasing, constant, and increasing hazard functions, respectively, the required number of events $(\tilde{n})$ calculated for a one-sided type I error rate 5% and power of 80% are given in Table 8.5. The distribution of $\log(T_2/T_1)$ does not depend on the correlation between the pair of TTPs; it depends on the shape parameter κ which plays a similar role as the correlation parameter ν for the GBVE model. The shape parameter $\kappa = 1$ and 2 in the Weibull frailty model correspond to the cases of independent ($\nu = 1$ or $\rho = 0$) and moderate correlation ($\rho \simeq 0.57$) for the GBVE model, respectively. Thus, study design is not very efficient under the Weibull frailty model when $\kappa \leq 1$ and more efficient when $\kappa > 1$ by using the score test Q (see Table 8.5).

8.6.5 Study Design Using Generalized Effect Size

In a real trial design, because the relapsed patients are identified and enrolled in the study at a closer time period, there is no historical data available to estimate the correlation ρ between the pair of (T_1, T_2) and to validate the underlying bivariate survival model. Therefore, an alternative way to design the study is using a generalized treatment effect size $p = S_{\mathrm{GMI}}(1) = P(T_2 > T_1)$, which has been considered for assessing treatment effects by O'Brien (1988) and Hauck et al. (2000). The generalized treatment effect size is intuitive to clinicians. It is the probability of T_2 greater than T_1. Under the null hypothesis (no treatment effect) $S_{\mathrm{GMI}}(1) = 50\%$, and one can set an alternative hypothesis of $S_{\mathrm{GMI}}(1) = 65\%$, 70%, or 75%, that is a 15%, 20% and 25% increase in the probability of TTP for the current therapy than that of TTP for the previous therapy. The required number of paired events calculated using formula (8.19) are 88, 50 or 32, respectively, for a one-sided type I error rate 5% and power of 80%. This design makes no assumption for the correlation and bivariate survival distribution. Of course, whether a 15% to 25% increase of the generalized effect size is achievable depends on how high the paired TTP are correlated and the structure of the bivariate distribution to connect the correlation and hazard ratio into the generalized effect size $S_{\mathrm{GMI}}(1)$. The relationship between an amount of increasing the probability $S_{\mathrm{GMI}}(1)$ and an amount of increasing the hazard ratio is illustrated in Figure 8.4 under the GBVE model with a moderate correlation $\rho = 0.5$. For example, the alternative hypotheses of $S_{\mathrm{GMI}}(1) = 65\%$, 70%, and 75% correspond to the hazard ratio of 0.708, 0.623 and 0.542, respectively.

8.7 Counting Censoring

8.7.1 Simple Adjustment

To calculate sample size (number of patients), we need to adjust the required number of events $\tilde{n}$ by a percentage of the cases in which T_{2i} is censored. In general, we recommend a 10% increase to the number of events for the total sample size, that is the sample size can be calculated by

$$n = \frac{\tilde{n}}{1 - 10\%} = 1.11\tilde{n}.$$

For the score test Q, if $T_{2i} \leq T_{1i}$ and T_{2i} is censored, the pair does not contribute to the test statistics and makes no contribution to the study power. Thus, we need only to adjust the required number of paired events $\tilde{n}$ by a percentage of the cases in which $T_{2i} \leq T_{1i}$ and T_{2i} is censored, which is often a small percentage. Thus, we recommend a 5% increase to the number of

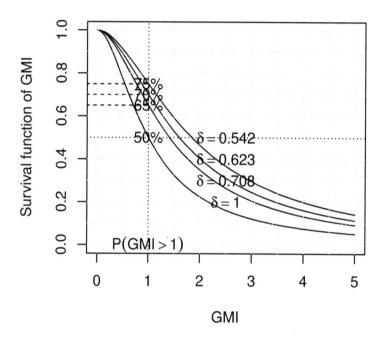

FIGURE 8.4: Graphic representation of the relationship between the survival distribution of GMI and generalized effect size $p = P(T_2 > T_1)$.

events for the total sample size, that is sample size can be calculated by

$$n = \frac{\tilde{n}}{1 - 5\%} = 1.05\tilde{n}.$$

Finally, the follow-up time should be longer than the median of TTP of T_2. For relapse or refractory cancer patients, the median TTP is often within one-year. Thus, a two-year follow-up could be sufficient for the trial to observe the planned number of events.

8.7.2 Uniform Accrual on Current Therapy

If we assume that patients on the current therapy were uniformly enrolled on interval $[0, t_a]$, followed for a period of t_f and no loss to follow-up, then the administrative censoring time C_2 for patients on current therapy is uniformly distributed on interval $[t_f, \tau]$, where $\tau = t_a + t_f$ is the total study duration. We further assume that TTP of previous therapy T_1 follows the Weibull distribution $S_1(t) = e^{-\lambda t^\kappa}$, which could be estimated from previous therapy TTP data. Then, the survival distribution of censoring time $G_{\text{GMI}}(u) = P(C_2/T_1 > u)$ can be derived as follows (see Appendix H):

$$G_{\text{GMI}}(u) = 1 - \frac{1}{t_a} \int_{t_f}^{\tau} e^{-\frac{\lambda}{u^\kappa} t^\kappa} dt, \quad u > 0.$$

Thus, the asymptotic variance $\sigma^2_{\text{GMI}}(q_0)$ can be calculated as follows:

$$\sigma^2_{\text{GMI}}(q_0) = \int_0^{q_0} \frac{\lambda_{\text{GMI}}(u)}{S_{\text{GMI}}(u) G_{\text{GMI}}(u)} du,$$

and sample size can be calculated using the formulae developed in previous sections under the assumption of the uniform censoring of TTP on the current therapy.

8.8 Simulation

To study the performance of the test statistic Q and accuracy of the formula (3) in finite sample cases, we conducted simulation studies under the GBVE model and Weibull-Gamma frailty model. Random bivariate paired samples of (T_{1i}, T_{2i}) are generated from GBVE$(\theta_1, \theta_2, \nu)$ according to the method given in Appendix G, and the Gamma frailty is generated from gamma distribution of Gamma$(2, 2)$. Without loss of generality, the parameter θ_1 is set to 1; inverse hazard ratio δ^{-1} set from 1.2 to 2.5 by 0.1; correlation ρ set from 0.1 to 0.8 by 0.1 for the GBVE model; R set from 1.5 to 2.5 by 0.1 and $\kappa = 0.5$, 1, and 2 for the Weibull-Gamma frailty model. The corresponding number of events are calculated by formula (8.19) with a type I error rate 5% (one-sided) and power of 80%. The simulations are done for the number of events, no censoring is assumed. The empirical type I error and empirical power are simulated based on 100,000 simulated trials (Tables 8.4 and 8.5). When the number of events is between 30 and 100, which is the size for the majority of phase II trials, the simulation results show that the test statistic Q preserves the type I error well and formula (8.19) provides correct power for the study design. When the number of events is small, the empirical type I error could be slightly inflated. This is because of the discreteness of the test statistic Q.

8.9 Example and R Code

Example 13 *Continuation of Example 12*

The data from a trial reported by Bonetti et al. (2001) is used to illustrate study design using proposed methods. In the trial, treatment efficacy was assessed by GMI for a second-line therapy (combination of LV-modulated 5-FU and oxaliplatin) in patients with advanced colon carcinoma after failure of a first-line chemotherapy (LV-modulated 5-FU or raltitrexed). A scatter plot of 34 observed pairs of TTPs (reconstructed data from Figure 1 of Kovalchik and Mietlowski, 2011) is given in Figure 8.1, which shows a significant correlation between the pairs (correlation $\rho = 0.514$). Sixteen patients (47%) had a GMI > 1.3, twenty-two patients (64.7%) had a GMI > 1 and three patients (9%) had censored T_2 less than T_1.

The marginal distributions of T_1 and T_2 are estimated by Kaplan-Meier curves and fitted by the Weibull models with a common shape parameter κ (Figure 8.5). The estimated scale parameters are $\hat{\theta}_1 = 21.06$ and $\hat{\theta}_2 = 31.89$, shape parameter is $\hat{\kappa} = 1.91$, and median ratio $\hat{\theta}_2/\hat{\theta}_1 = 1.514$. Figure 8.5 shows that the Weibull models fit the data well. The Kaplan-Meier curve and fitted log-logistic distribution of the GMI is given in Figure 8.6. An estimate of the generalized effect size $p = S_{GMI}(1)$ from the Kaplan-Meier curve is 72% (95% confidence interval: 53%-84%). Suppose we want to design a new trial under the frailty model to detect a median ratio $R = 1.5$ with shape parameter $\kappa = 1.9$, or $p = S_{GMI}(1) = 68\%$. Given one-sided type I error rate 5% and power of 80%, the required number of events calculated by formula (8.19) is $\tilde{n} = 61$. In case the Weibull frailty model can't be justified, the trial can be designed by using the generalized effect size $p = S_{GMI}(1)$. For example, the generalized effect size at alternative hypothesis is set to 70%, the required number of events is $\tilde{n} = 50$. By adjusting 10% censored observations which do not contribute to the test statistic, a total of $n = 68$ and $n = 56$ patients are needed for the trial corresponding to the above two design scenarios, respectively.

```
#### Sample size calculation under GBVE model ######
```

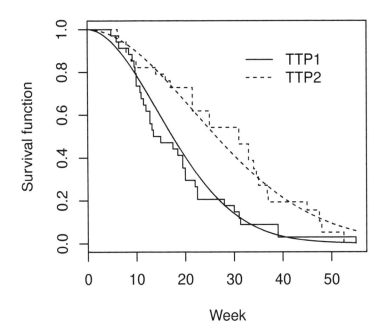

FIGURE 8.5: Kaplan-Meier curves (step functions) and fitted Weibull distributions (smooth curves) for first-line therapy arm and second-line therapy arm, respectively.

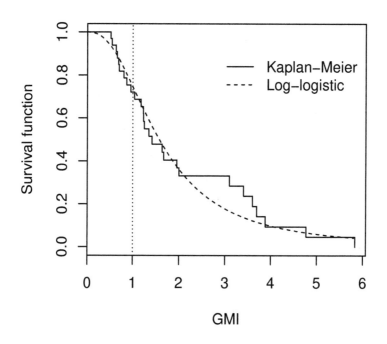

FIGURE 8.6: Kaplan-Meier curve (step function) and fitted log-logistic distribution.

```
Size=function(rho, delta, alpha, beta)
{g=function(nu){2*gamma(nu+1)^2/(gamma(2*nu+1))-1-rho}
 nu=uniroot(g, lower=0, upper=1)$root
 p=1/(1+delta^(1/nu))
 q0=qchisq(1-alpha,1)
 f=function(ncp){pchisq(q0,1,ncp)-beta}
 ncp=uniroot(f, lower=1, upper=20)$root
 n=ncp/(4*(p-0.5)^2)
 ans=ceiling(n)
 return(ans)
}
Size(rho=0.5, delta=0.6, alpha=0.05, beta=0.2)
43
#### Sample size calculation under Gamma Frailty model ######
Size=function(kappa, R, alpha, beta)
{p=1/(1+R^(-kappa))
 q0=qchisq(1-alpha,1)
 f=function(ncp){pchisq(q0,1,ncp)-beta}
 ncp=uniroot(f, lower=1, upper=20)$root
 n=ncp/(4*(p-0.5)^2)
 ans=ceiling(n)
 return(ans)
}
Size(kappa=1, R=2, alpha=0.05, beta=0.2)
[1] 71
#### Sample size using generalized effect size p ######
Size=function(p, alpha, beta)
{f=function(ncp){pchisq(q0,1,ncp)-beta}
 q0=qchisq(1-alpha,1)
 ncp=uniroot(f, lower=1, upper=20)$root
 n=ncp/(4*(p-0.5)^2)
 print(round(n))
}
Size(p=0.70, alpha=0.05, beta=0.2)
49
########### Score test statistics ######
Test=function(data)
{TTP1=data[,2]
 TTP2=data[,4]
 status=data[,3]
 GMI <- TTP2/TTP1
 n=nrow(data)
 r=NULL
 for (i in 1:n)
  {
   if (GMI[i]>1) {r[i]=1}
   if (GMI[i]<=1 & status[i]!=0) {r[i]=-1}
   if (GMI[i]<=1 & status[i]==0) {r[i]=0}
  }
 stat=sum(r)^2/sum(r^2)
 p=1-pchisq(stat, 1)
 ans=round(p,5)
 return(ans)
}
```

8.10   Discussion

The GMI as a primary endpoint for single-arm phase II trials with cytostatic cancer treatment was first proposed by Von-Hoff (1998). The GMI has advantages of being a more suitable indicator of delayed growth than traditional response measures. Furthermore, with the GMI endpoint, patients served as their own controls, which makes the outcome less heterogeneous and more personalized, thus more feasible to establish benchmarks for evidence of treatment efficacy (Kovalchik and Mietlowski, 2011). However, it has not been put into practice because of lack of sound statistical methods for the trial design. Even though, Mick et al. (2000) have discussed a hypothesis testing approach and Kovalchik and Mietlowski (2011) recommended an estimation approach under the Weibull frailty model, their results are limited to the simulations under the restricted model assumption. Directly using their simulation results could be misleading to the trial design.

Von-Hoff considered GMI > 1.3 as a criterion for treatment activity. Under the $\text{GBVE}(\theta_1, \theta_2, \nu)$ model, the survival function of GMI can be calculated as

$$S_{\text{GMI}}(s) = P(\text{GMI} > s) = \frac{1}{1 + (s\delta)^{1/\nu}}.$$

Thus, $S_{\text{GMI}}(1.3) = 1/\{1 + (1.3\delta)^{1/\nu}\}$. Von-Hoff et al. (2010) set the null hypothesis $S_{\text{GMI}}(1.3) = 15\%$ vs. the alternative hypothesis $S_{\text{GMI}}(1.3) = 30\%$ for the study design. However, the choice of $S_{\text{GMI}}(1.3) = 15\%$ as an inactive criteria and $S_{\text{GMI}}(1.3) = 30\%$ as an active criteria are arbitrary and difficult to justify. For example, under the GBVE model with a moderate correlation $\rho = 0.5$, $S_{\text{GMI}}(1.3) = 15\%$ vs. 30% correspond to the hazard ratio 2.04 vs. 1.23, respectively, which obviously were both set too low for the hypothesis of treatment efficacy.

To overcome those drawbacks, a sample size formula is derived under an AFT model for the paired TTP data. Trial design with GMI endpoint was discussed under a bivariate exponential model, Weibull frailty model and generalized treatment effect size. The advantage of using bivariate exponential model or Weibull frailty model is that the sample size can be calculated in a traditional way through specifying the hazard ratio of the paired TTPs. However, because the relapsed patients are identified and enrolled in the study at a closer time period, there is no historical data available to estimate the correlation ρ between the paired TTPs and to validate the underlying bivariate survival model. Thus, an alternative design using the generalized effect size is attractive which does not rely on specifying the correlation and bivariate survival distribution. The number of events or sample size calculation is extremely simple by using formula (8.19).

Finally, since in relapse and refractory cancer patients, successive TTPs tend to be shorter and shorter, Kovalchik and Mietlowski (2011) discussed a trial design under the null hypothesis of median GMI $= 0.7$ vs. alternative hypothesis median GMI $= 1$ as the treatment equivalence or Von-Hoff's median GMI $= 1.3$ as the treatment efficacy. In this case, the generalized effect size can be defined by $p = S_{\mathrm{GMI}}(0.7)$ and the null hypothesis will be $S_{\mathrm{GMI}}(0.7) = 50\%$ vs. alternative, of say $S_{\mathrm{GMI}}(0.7) = 70\%$. The sample size formula (8.19) can still be used for such trial design.

Overall, the derived formula provides accurate estimation of the number of events. The trial design using the generalized effect size is simple and intuitive to clinicians. The proposed study designs provide sound statistical methods for single-arm phase II trials to test more appropriate endpoints such as paired TTPs for new cancer treatment paradigms being advanced to clinical trials for relapse or refractory cancer patients.

9

Bayesian Single-Arm Phase II Trial Design

9.1 Introduction

For cancer clinical trials on immunotherapy and molecularly targeted therapy, time-to-event endpoints, such as progression-free survival is often a desired endpoint. Single-arm phase II trial designs with time-to-event endpoints have been largely studied by frequentist approaches prospectively, whereas Bayesian counterparts of such trial designs are limited. Thall et al. (2005) proposed a Bayesian phase II design based on the posterior probability of mean survival time. Zhao et al. (2012) developed a Bayesian decision theoretic two-stage phase II survival trial. Shi and Yin (2018) proposed a Bayesian enhancement two-stage design for a single-arm phase II trial with time-to-event endpoints. Yuan et al. (2016) proposed a practical Bayesian design for platform trials with time-to-event endpoints. Lin et al. (2020) presented a time-to-event Bayesian optimal phase II trial design for cancer immunotherapy. Also, Zhou et al. (2020) recently developed a Bayesian multi-stage phase II trial design with a frequentist type I error control. One advantage of Bayesian methods is that study designs can be derived from the exact posterior distributions that accommodate the single-arm phase II trial. For example, most pediatric cancers are rare and have small sample sizes, and frequentist designs based on large sample asymptotic distributions may not be appropriate for small sample sizes.

However, all the above methods rely on the assumption that the time-to-event follows a parametric model (e.g., an exponential or Weibull distribution) and sample size calculation requires the assumption of uniform accrual and censoring distributions. However, in practice, it is extremely challenging to make an assumption for the accrual distribution, as the accrual rate cannot be controlled and the distribution of the censoring time remains unknown. All the above issues will result in inaccurate estimation of sample size, which elevates the risk of drug development. To overcome these issues, especially parametric assumption for the time-to-event, we adapt a transformed time-to-event method used by Cotterill and Whitehead (2015), which is based on the proportional hazards assumption in contrast to existing methods that adopt parametric models. Using this transformed time-to-event method, we propose a Bayesian one-stage design and a two-stage design that are driven by the number of events instead of the number of patients. As a result, the proposed designs are independent of the accrual and censoring distributions. The advantage of such distribution-free designs is to avoid possible misspecifications of the time-to-event, accrual, and censoring distribution assumptions that are often very difficult to be specified accurately prior to the study.

9.2   Transformed Time-to-Event Model

Suppose that during the accrual phase of a trial with a new treatment, n patients are enrolled in the study. We assume that $\{T_i, i = 1, \ldots, n\}$ are independent failure times with a common survival distribution, $S(t)$ and $\{C_i, i = 1, \ldots, n\}$ are independent censoring times, and $\{T_i\}$ and $\{C_i\}$ are independent. Then, the observed failure times X_i and failure indicator Δ_i are

$$X_i = \min(T_i, C_i) \quad \text{and} \quad \Delta_i = I(T_i \leq C_i), \quad i = 1, \cdots, n,$$

where $I(\cdot)$ is an indicator function. Let $S_0(t)$ be a known survival distribution of a reference group. Our aim is to test the following hypothesis:

$$H_0 : S(t) \leq S_0(t) \quad \text{vs} \quad H_1 : S(t) > S_0(t) \quad \text{for all } t > 0. \qquad (9.1)$$

Instead of assuming a distribution for the failure time, we assume a proportional hazards model

$$S(t) = [S_0(t)]^\delta,$$

where δ is the hazard ratio of the treatment group vs. reference group. Then, the hypothesis (9.1) is equivalent to the following hypothesis:

$$H_0 : \delta \geq 1 \quad \text{vs} \quad H_1 : \delta < 1. \qquad (9.2)$$

By using the transformed observations for testing the hypothesis (9.2), let $\tilde{T}_i = \Lambda_0(T_i) = -\frac{1}{\delta} \log S(T_i)$, where $\Lambda_0(t) = -\log S_0(t)$ and $S(T_i)$ is uniform$(0, 1)$; thus, $\tilde{T}_i$ follows an exponential distribution with rate parameter δ. Let $W_i = \min(\tilde{T}_i, \tilde{C}_i) = \Lambda_0(X_i)$, where $\tilde{C}_i = \Lambda_0(C_i)$. We obtain the transformed data $\{W_i, \Delta_i, i = 1, \ldots, n\}$, where $\Delta_i = I(T_i \leq C_i) = I(\tilde{T}_i \leq \tilde{C}_i)$. The likelihood function of the transformed data is given by

$$L = \prod_{i=1}^{n} f(W_i)^{\Delta_i} S(W_i)^{1-\Delta_i} \propto \prod_{i=1}^{n} \delta^{\Delta_i} e^{-\delta W_i} = \delta^d e^{-\delta U},$$

where $d = \sum_{i=1}^{n} \Delta_i$ is the number of events and $U = \sum_{i=1}^{n} W_i$ is the total observation time (after transformation). Since (d, U) is a sufficient statistic, we can write the observed data by $D = (d, U)$.

9.3 Bayesian One-Stage Design

Cotterill and Whitehead (2015) used a gamma prior distribution on δ. Specifically, let $\delta \sim \text{Gamma}(a, b)$ with density function $\pi(\delta) \propto \delta^{a-1}e^{-b\delta}$, where a is a shape parameter and b is a rate parameter. Increasing values of a correspond to increasing levels of certainty in the prior distribution of δ if b are given. Due to the conjugacy relationship between the gamma distribution and the likelihood function of δ, the posterior distribution of δ follows $\pi(\delta|D) \propto \delta^{a+d-1}e^{-(b+U)\delta}$, which is again a gamma distribution $\text{Gamma}(a + d, b + U)$. For testing the hypothesis (9.2) after obtaining the number of events $d = m$ and total observation time U and its critical value k, from the perspective of Bayesian analysis, there is convincing evidence that the new treatment is promising if the posterior probability $P(\delta < 1|m, U)$ is no less than η when rejecting the null $U \geq k$, where η is a value chosen to be close to 1. More specifically, we require that the posterior probability satisfies the following "Go" criterion:

$$\inf_{U \geq k} P(\delta < 1|m, U) \geq \eta.$$

It can be shown that the posterior probability $P(\delta < 1|m, U)$ is a monotone increasing function in U, given the number of events $d = m$ and the Gamma prior parameters a and rate b (Appendix I). Then, we have

$$\inf_{U \geq k} P(\delta < 1|m, U) = P(\delta < 1|m, k) \geq \eta. \tag{9.3}$$

We declare that the new treatment is promising if $U \geq k$. The posterior probability $P(\delta < 1|m, k)$ under the Gamma posterior distribution can be calculated by the numerical integration

$$P(\delta < 1|m, k) = \int_0^1 \frac{(b+k)^{a+m}}{\Gamma(a+m)} t^{a+m-1} e^{-(b+k)t} dt.$$

To determine the number of events and critical value k for the reject region, given a hazard ratio δ_1 (< 1) in the alternative hypothesis, we require that the posterior probability $P(\delta > \delta_1|m, U)$ in the accepting region satisfies the following "no-Go" criterion:

$$\inf_{U < k} P(\delta > \delta_1|m, U) = P(\delta > \delta_1|m, k) \geq \zeta, \tag{9.4}$$

where ζ is a value close to 1. This is because a large posterior probability $P(\delta > \delta_1|m, U)$ provides convincing evidence that the new treatment is unpromising. Thus, we will reject the null hypothesis and abandon the new treatment if $U < k$.

However, there is no explicit solution for the number of events m and critical value k under the Gamma posterior distribution. Therefore, the following search algorithm 1 is proposed to determine the number of events $d = m$, and the critical value k of the total observation time U to satisfy the equations of (9.3) and (9.4).

Algorithm 1

1. Start with minimum number of events $m = 1$ up to maximum number of events $m_{max} = 100$.

2. Given $m = i$, find a minimum total of observation time (after transformation) k_0 such that $P(\delta < 1|m, k_0) = \eta$, and calculate the corresponding $P(\delta > \delta_1|m, k_0)$.

3. If $P(\delta > \delta_1|m, k_0) \geq \zeta$, we stop the algorithm and set $m = i$ and $k = k_0$; otherwise, we set $m = i + 1$ and repeat step 2.

The above proposed Bayesian one-stage design can be executed as follows in practice: enroll patients until m events are observed, and then calculate the transformed total observation time U. If $U \geq k$, we declare that the new treatment is promising and warrants investing in a large-scale phase III trial. If not, we consider that the new treatment is unpromising.

To control the frequentist type I error rate (α) and power $(1 - \beta)$ for the proposed Bayesian design, we derive theoretical relationships between the α, β and posterior probability thresholds η, ζ (Appendix J) as follows:

$$P(\delta < 1|m, m + \sqrt{m}z_{1-\alpha}) = \eta, \qquad (9.5)$$

$$P(\delta > \delta_1|m, \sqrt{m}(z_{1-\alpha} + z_{1-\beta})/(1 - \delta_1)) = \zeta. \qquad (9.6)$$

Thus, to obtain the desired frequentist type I error rate α and power $1 - \beta$, we can use the above two equations to calibrate η and ζ to achieve desirable operating characteristics for the Bayesian design.

9.4 Bayesian Two-Stage Design

To stop the trial early if the new treatment shows no sign of being promising, we propose a flexible Bayesian two-stage design. Specifically, at the first stage, based on the observed number of events $d_1 = m_1$ and a critical value k_1 of the total observation time U_1, we make a decision of stopping for futility or continuation. The posterior probability $P(\delta > \delta_1|m_1, U_1)$ in the accepting region should satisfy the following futility stopping criterion:

$$\inf_{U_1 < k_1} P(\delta > \delta_1|m_1, U_1) = P(\delta > \delta_1|m_1, k_1) \geq \zeta_1, \qquad (9.7)$$

where $\zeta_1 = \eta(1 - \xi\tilde{t})$ with $0 < \xi < 1$ and $\tilde{t} = m_1/m$ is the information fraction at the first interim analysis. Note that ζ_1 is decreasing with $\tilde{t}$, which reflects the clinical rationale that at the first stage in which there are fewer events, it would be prudent and conservative to terminate a study early for futility.

At the second stage, depending on the observed number of events $d = m$ and critical value k of the total observation time U, we require that the posterior probability $P(\delta > \delta_1|m, U)$ in the accepting region satisfies the following "no-Go" criterion:

$$\inf_{U < k} P(\delta > \delta_1|m, U) = P(\delta > \delta_1|m, k) \geq \zeta, \tag{9.8}$$

where $\zeta = \eta(1 - \xi)$ since $\tilde{t} = 1$ now. Meanwhile, we also require that the posterior probability $P(\delta < 1|m, U)$ in the rejection region satisfies the following "Go" criterion:

$$\inf_{U \geq k} P(\delta < 1|m, U) = P(\delta < 1|m, k) \geq \eta. \tag{9.9}$$

For the final data analysis, we reject the null hypothesis if $U \geq k$.

To determine the two-stage design parameters (m_1, k_1, m, k), we prespecify an information fraction $\tilde{t} = m_1/m$ for an interim analysis. The information fraction determines the number of events $m_1 = m\tilde{t}$ for the first stage. We set the tuning parameter ξ to determine the first and second stage posterior probability thresholds $\zeta_1 = \eta(1 - \xi\tilde{t})$ and $\zeta = \eta(1 - \xi)$ such that the trial has desirable size and properties. The two-stage design algorithm 2 is described as follows:

Algorithm 2

1. Start with a minimum number of events $m = 1$ up to a maximum number of events $m_{max} = 100$.

2. Given $m = i$, find a minimum total of observed survival time (after transformation) k_0 such that $P(\delta < 1|m, k_0) = \eta$.

3. Calculate the $P(\delta > \delta_1|m, k_0)$. If $P(\delta > \delta_1|m, k_0) \geq \zeta$, where $\zeta = \eta(1 - \xi)$. We set $m = i$ and $k = k_0$; otherwise, we set $m = i + 1$ and repeat step 2.

4. Calculate $m_1 = m\tilde{t}$ and solve the total observation time k_1 such that $P(\delta > \delta_1|m_1, k_1) = \zeta_1$, where $\zeta_1 = \eta(1 - \xi\tilde{t})$.

The flexible Bayesian two-stage design can be executed as follows. Enroll patients until m_1 events are observed. Then, calculate the transformed total observation time U_1 at the first stage. If $U_1 > k_1$, the trial goes to the second stage, otherwise the trial stops for futility. At the second stage, enroll patients until m events are observed, and then calculate the transformed total observation time U. If $U > k$, we declare that the new treatment is promising and warrants further study in a large-scale phase III trial; otherwise, the new treatment is not promising and is not worth further study.

9.5   Simulation

In this section, we illustrate the proposed Bayesian one-stage and two-stage designs and conduct simulations to study their frequentist properties, assuming that the 3-year survival probability for the reference group is $S_0(3) = 0.53$. The trial is designed to detect hazard ratios of $\delta_1 = 0.6, 0.7$, or a negative log hazard ratio $\theta_1 = -\log(\delta_1)$. We take the prior distribution of δ to be a Gamma(a, b) distribution with shape parameters $a = 2, 5, 10$ and rate parameter $b = 2a/(1 + \delta_1)$, and the corresponding posterior mean is $a/b = 0.8$ and standard deviation is $\sqrt{a}/b = 0.8/\sqrt{a}$. Thus, with a small value of a, the standard deviation is large and the prior is a weak or non-informative prior, and with a large value of a, the standard deviation is small and the prior is more informative. As derived in Section 3.1, the posterior distribution of δ is a Gamma$(a + d, b + U)$, where d and U are the total number of events and total observation times, respectively (after transformation).

For one-stage designs, to be confident in the conclusion drawn from the trial, we choose $\eta = 0.95, 0.9$ and $\zeta = 0.9, 0.85, 0.8$. As the cost of falsely stopping for futility is lower than that for falsely continuing the trial with an inefficacious treatment into further large-scale studies, we typically set $\eta \geq \zeta$.[12] The required number of events m and critical value k are calculated using searching algorithm 1. The results are presented in Table 9.1. For example, with $\eta = 0.9$ and $\zeta = 0.8$, and a weak prior $a = 2$, the required number of events is $m = 16$ and critical values $k = 21.11$, respectively. We can see that two methods give almost identical designs. Similar results are also seen for other design scenarios. We can also see that with a more informative prior, the required number of events decreases. For example, given that $\delta = 0.6$, $\eta = 0.9$, and $\zeta = 0.85$, an increase in a from 2 to 10 indicates that the prior mean hazard rates are all 0.8. Whereas the prior standard deviation of hazard rates decreases from 0.57 to 0.25, the required number of events decreases from 19 to 11.

We conducted simulations to study frequentist properties of the proposed Bayesian designs. For each design scenario, 10,000 simulated trials (under exponential distribution) were used to estimate the frequentist empirical type I error rate and power. The frequentist type I error rate and power depend on the level of posterior probability thresholds η and ζ. Both simulations and numerical calculations using equations (9.5) and (9.6) showed that the frequentist type I error rate α and power $1 - \beta$ are much closer to those of posterior probability thresholds $1 - \eta$ and ζ, respectively, for one-stage design with a weak or non-informative prior. The simulation results also showed that the frequentist type I error rate inflated and power decreased a bit as the prior became more informative if the same set of thresholds of η and ζ were applied under a given alternative δ. For instance, given $\delta = 0.6$, with $\eta = 0.9$ and $\zeta = 0.8$, as a increases from 2 to 10, the type I error rate inflated from

0.109 to 0.140 and the power decreased from 0.795 to 0.644. This finding is expected since the same set of thresholds were used through various priors. On the other hand, this finding also informs statisticians of the importance of calibrating the thresholds of η and ζ. For example, if we still want to use $a = 10$ with $\delta = 0.6$ to control the type I error rate under 10% and have a power of at least 80%, from Table 9.1, we can set $\eta = 0.95$ and $\zeta = 0.90$. Then, the type I error rate is 7.2% and the power is 84%. This set of η and ζ over-controls the type I error rate, but showing an example and the proposed searching algorithm can be used for finding the η and ζ to achieve the target type I error rate and power goal.

For two-stage designs, we choose $\eta = 0.9, 0.95$ and the tuning parameter $\xi = 0.01$. We solve the number of events m and critical value k for the second stage from equations (9.8) and (9.9) by using the searching algorithm 2. Then, given the interim information fraction $\tilde{t} = 0.4, 0.5, 0.6$, we calculate the number of events $m_1 = m\tilde{t}$ for the first stage and solve the critical value k_1 by using equation (9.7). The study designs obtained for these three scenarios are given in Table 9.2. We conducted simulations to study frequentist properties of the proposed flexible Bayesian two-stage designs. For each design scenario, 10,000 simulated trials (under exponential distribution) were generated to estimate the empirical type I error rate and power, probability of early stopping (PET) at null and alternative, and expected sample size (ES) under the null and alternative. With $\eta = 0.9$, $\xi = 0.01$ and a weak prior $a = 2$, the type I error rate can be controlled below 10% whereas the power can be kept above 80%. With $\eta = 0.95$, $\xi = 0.01$, and a weak prior $a = 2$, the type I error rate can be kept below 5% whereas the power can be kept above 90%. This is possible because of the relationship between the frequentist type I error rate α and power $1 - \beta$ and the posterior probability thresholds η and ζ as derived for the one-stage design. For the two-stage design with the futility rule only for an interim analysis, the frequentist type I error rate and power are slightly reduced. For a more informative prior, the power is decreased and type I error rate is increased. The probabilities of early stopping were below 20% and above 60% under the null and alternative, respectively. The ES (expected sample size) under the alternative did not reduce much from the total sample size; however, the ES under the null were reduced more due to the PET (high probability of early stopping). Furthermore, we explored the sensitivity of the proposed Bayesian design compared to those of prior distributions. Results showed that the proposed Bayesian designs are quite robust and stable for the non-informative prior (see Table 9.3). Therefore, we recommend that the non-informative prior be used in real practice for keeping the study designs with a controlled type I error and high power.

In practice, the number of patients or the study duration may also be important information for planning a trial. R codes are provided in the next section for convenient use of the proposed methods for the event-driven design as well as sample size calculation under the four parametric distributions Weibull, gamma, log-normal and log-logistic.

9.6 Discussion

The existing frequentist methods for designing single-arm phase II trials with time-to-event endpoints are based on large sample asymptotic distributions that may not be appropriate when the sample size is small for single-arm phase II trials. Bayesian methods can provide exact posterior distributions by taking advantage of conjugate priors. However, due to the complexity caused by censoring, all these existing methods assumed that the time-to-event follows an exponential or Weibull distribution and focused on estimating the required sample size; however, sample size calculation needs the assumption of accrual and censoring distributions. Nevertheless, in practice, it is extremely difficult to make an accurate assumption for the accrual distribution due to the uncontrollable accrual rate. In addition, the distribution of censoring time is also unknown. All the above factors will compromise the performances of existing approaches. To overcome these difficulties, we present event-driven approaches for both one-stage and two-stage designs. The event-driven designs do not require knowledge of underlying survival, accrual, and censoring distributions; therefore, the proposed methods take the advantage of avoiding potential misspecification for the trial design.

The proposed Bayesian two-stage design provides a counterpart of Simon's two-stage design that allows early futility stopping when the new treatment shows no signs of efficacy. However, the proposed Bayesian designs take the advantage of using time-to-event endpoints, which has become very popular for cancer immunotherapy and targeted therapy trials. Furthermore, using the event-driven approach, study designs are robust and without assumptions of the underlying survival, accrual, and censoring distributions. Finally, Bayesian designs based on the exact posterior distributions accommodate single-arm phase II trials with small sample sizes. Therefore, the proposed event-driven Bayesian approaches can be considered as a useful tool for designing single-arm phase II cancer immunotherapy trials and rare cancer trials such as pediatric cancer studies.

9.7 R Code

This section provides R codes for Bayesian one-stage design 'BSize', for flexible Bayesian two-stage design 'B2Size' and Bayesian two-stage design with frequentist type I error rate and power targets 'optimal'.

```
#################### Bayesian One-Stage Design  ##########################
### Shape is shape parameter of the four parametric distributions; S0 is  #####
### the survival probability at x; ta and tf are accrual and follow-up    #####
### duration; delta is hazard ratio; eta and zeta are the Bayesian        #####
```

TABLE 9.1: Bayesian single-stage designs under various scenarios with confidence levels $\eta = 0.9, 0.95$ and $\zeta = 0.8, 0.85, 0.9$, hazard ratio $\delta = 0.6, 0.7$. Sample sizes were calculated under exponential survival distributions: Frequentist empirical type I error ($\hat{\alpha}$) and empirical power (EP) were estimated from 10,000 simulated trials under exponential distribution.

Parameters						One-stage design		
a	δ	η	ζ	m	k	n	$\hat{\alpha}$	EP
2	.6	.90	.80	16	21.11	41	.109	.795
5	.6	.90	.80	13	17.36	34	.122	.750
10	.6	.90	.80	8	11.11	21	.140	.644
2	.6	.90	.85	19	24.55	49	.109	.840
5	.6	.90	.85	16	20.80	41	.121	.808
10	.6	.90	.85	11	14.55	29	.144	.738
2	.6	.95	.85	25	33.58	64	.055	.837
5	.6	.95	.85	22	29.83	57	.061	.812
10	.6	.95	.85	17	23.58	44	.069	.748
2	.6	.95	.90	31	40.48	80	.053	.897
5	.6	.95	.90	28	36.73	72	.058	.878
10	.6	.95	.90	23	30.48	59	.072	.844
2	.7	.90	.80	33	40.26	76	.108	.797
5	.7	.90	.80	30	36.51	69	.119	.790
10	.7	.90	.80	25	30.26	57	.148	.776
2	.7	.90	.85	41	49.09	94	.112	.858
5	.7	.90	.85	38	45.34	87	.123	.850
10	.7	.90	.85	33	39.09	76	.144	.843
2	.7	.95	.85	54	66.35	124	.056	.855
5	.7	.95	.85	51	62.60	117	.062	.843
10	.7	.95	.85	46	56.35	105	.072	.832
2	.7	.95	.90	65	78.51	149	.056	.900
5	.7	.95	.90	62	74.76	142	.061	.896
10	.7	.95	.90	57	68.51	130	.070	.889

TABLE 9.2: Flexible Bayesian two-stage designs under various scenarios with prior parameter $a = 2, 5, 10$, information fraction $\tilde{t} = 0.4, 0.5, 0.6$, and tuning parameters $\eta = 0.9, 0.95$ and $\xi = 0.01$. Empirical type I error ($\hat{\alpha}$) and power (EP), early stopping probabilities under null (PET$_0$) and alternative (PET$_1$), expected sample size under null (ES$_0$) and alternative (ES$_1$) were estimated from 10,000 simulation runs under exponential distribution.

Design parameters					Stage I			Stage II			Frequentist operating characteristics					
η	a	t	ζ_1	ζ	m_1	n_1	k_1	m	n	k	$\hat{\alpha}$	EP	PET$_1$	ES$_1$	PET$_0$	ES$_0$
.9	2	.4	.896	.891	10	26	10.637	23	59	29.08	.082	.812	.122	54.97	.632	38.14
	2	.5	.896	.891	12	31	13.404	23	59	29.08	.081	.818	.121	55.61	.697	39.48
	2	.6	.895	.891	14	36	16.217	23	59	29.08	.082	.828	.119	56.26	.755	41.63
	5	.4	.896	.891	8	21	8.253	20	52	25.33	.094	.787	.134	47.85	.590	33.71
	5	.5	.896	.891	10	26	11.043	20	52	25.33	.093	.791	.141	48.33	.678	34.37
	5	.6	.895	.891	12	31	13.875	20	52	25.33	.092	.798	.143	49.00	.740	36.46
	10	.4	.896	.891	6	16	6.165	15	39	19.08	.110	.727	.176	34.95	.592	25.38
	10	.5	.896	.891	8	21	9.012	15	39	19.08	.109	.732	.187	35.63	.684	26.69
	10	.6	.895	.891	9	23	10.462	15	39	19.08	.108	.735	.193	35.91	.724	27.42
.95	2	.4	.946	.940	16	41	17.080	38	97	48.44	.044	.902	.058	93.75	.642	61.05
	2	.5	.945	.940	19	49	21.213	38	97	48.44	.042	.904	.061	94.07	.723	62.30
	2	.6	.944	.940	23	59	26.817	38	97	48.44	.043	.908	.062	94.64	.799	66.64
	5	.4	.946	.940	14	36	14.683	35	90	44.69	.048	.888	.064	86.54	.612	56.95
	5	.5	.945	.940	18	46	20.225	35	90	44.69	.047	.889	.070	86.92	.728	57.97
	5	.6	.944	.940	21	54	24.470	35	90	44.69	.046	.892	.072	87.41	.789	61.60
	10	.4	.946	.940	12	31	12.539	30	77	38.44	.057	.856	.085	73.09	.610	48.94
	10	.5	.945	.940	15	39	16.761	30	77	38.44	.057	.860	.089	73.62	.708	50.10
	10	.6	.944	.940	18	46	21.040	30	77	38.44	.056	.866	.090	74.21	.780	52.82

TABLE 9.3: Flexible Bayesian two-stage designs under various scenarios with prior parameter $a = 0.01, 0.1, 1$, information fraction $\tilde{t} = 0.4, 0.5, 0.6$, and tuning parameters $\eta = 0.9, 0.95$ and $\xi = 0.01$. Empirical type I error ($\hat{\alpha}$) and power (EP), early stopping probabilities under null (PET$_0$) and alternative (PET$_1$), expected sample size under null (ES$_0$) and alternative (ES$_1$) were estimated from 10,000 simulation runs under exponential distribution.

		Design parameters			Stage I			Stage II			Frequentist operating characteristics					
η	a	$\tilde{t}$	ζ_1	ζ	m_1	n_1	k_1	m	n	k	$\hat{\alpha}$	EP	PET$_1$	ES$_1$	PET$_0$	ES$_0$
.9	0.01	.4	.896	.891	10	26	10.447	25	64	31.58	.075	.826	.113	59.71	.610	40.82
	0.01	.5	.896	.891	13	34	14.527	25	64	31.58	.075	.838	.110	60.70	.701	42.97
	0.01	.6	.895	.891	15	39	17.319	25	64	31.58	.074	.844	.108	61.30	.753	45.17
	0.10	.4	.896	.891	10	26	10.455	25	64	31.57	.075	.826	.113	59.71	.611	40.78
	0.10	.5	.896	.891	13	34	14.538	25	64	31.57	.075	.838	.110	60.70	.702	42.94
	0.10	.6	.895	.891	15	39	17.332	25	64	31.57	.074	.843	.108	61.30	.754	45.15
	1.00	.4	.896	.891	10	26	10.534	24	62	30.33	.104	.842	.117	57.79	.620	39.68
	1.00	.5	.896	.891	12	31	13.275	24	62	30.33	.101	.848	.116	58.40	.685	40.77
	1.00	.6	.895	.891	15	39	17.467	24	62	30.33	.098	.857	.113	59.40	.763	44.45
.95	0.01	.4	.946	.940	16	41	16.902	40	103	50.94	.039	.906	.055	99.59	.626	64.19
	0.01	.5	.945	.940	20	52	22.344	40	103	50.94	.038	.910	.057	100.09	.727	65.92
	0.01	.6	.944	.940	24	62	27.921	40	103	50.94	.038	.914	.056	100.70	.802	70.12
	0.10	.4	.946	.940	16	41	16.909	40	103	50.93	.040	.906	.055	99.59	.627	64.13
	0.10	.5	.945	.940	20	52	22.355	40	103	50.93	.038	.910	.057	100.09	.728	65.87
	0.10	.6	.944	.940	24	62	27.934	40	103	50.93	.038	.914	.056	100.70	.802	70.12
	1.00	.4	.946	.940	16	41	16.986	39	100	49.69	.044	.904	.056	96.70	.635	62.53
	1.00	.5	.945	.940	20	52	22.463	39	100	49.69	.042	.908	.059	97.17	.735	64.72
	1.00	.6	.944	.940	24	62	28.067	39	100	49.69	.042	.912	.059	97.76	.808	69.30

```
### posterior confidence level for "GO" and "no-Go" criterions; zeta1 is  #####
### the Bayesian posterior confidence level for stop futility; emax is     #####
### the max number of events for iteration of the searching algorithm;      #####
#### dist is option for set up reference survival dist as one of the four #####
###  following  dist. including Weibull (WB), log-normal(LN), gamma (GM),  #####
###  log-logistic (LG).                                                     #####
################################################################################
BSize=function(shape,S0,x,ta,tf,a,delta,eta,zeta,emax,dist)
{
    S1=S0^delta
    tau=ta+tf
    if (dist=="WB"){
     f=function(a,b,u){dweibull(u,a,b)}
     scale1=x/(-log(S1))^(1/shape)
    }

    if (dist=="LN"){
     f=function(a,b,u){dlnorm(u,b,a)}
     scale1=log(x)-shape*qnorm(1-S1)
    }

    if (dist=="LG"){
     f=function(a,b,u){(a/b)*(u/b)^(a-1)/(1+(u/b)^a)^2}
     scale1=x/(1/S1-1)^(1/shape)
    }

    if (dist=="GM"){
     s=function(a,b,u){1-pgamma(u,a,b)}
     f=function(a,b,u){dgamma(u,a,b)}
     root1=function(t){s(shape,t,x)-S1}
     scale1=uniroot(root1,c(0,10))$root
    }
    G=function(u){1-punif(u, tf, tau)}
    g=function(u){f(shape,scale1,u)*G(u)}
    p=integrate(g, 0, tau)$value

    b=2*a/(1+delta)
    for (s in 1:emax){
    f=function(k){
     pgamma(1,shape=a+s,rate=b+k)-eta
    }
    k=uniroot(f,c(0,100))$root
    m=s; k=round(k,2)
    exp=pgamma(delta,shape=a+s,rate=b+k)
    if (exp<=1-zeta) break
  }
 n=ceiling(m/p)
 return(c(a,eta,zeta,m,n,k))
}
BSize(shape=1,S0=0.53,x=3,ta=4,tf=2,a=2,delta=0.6,eta=0.95,
      zeta=0.85,emax=100,dist="WB")
2 0.95 0.85    25    64 33.58
################# Flexible Bayesian Two-Stage Design ########################
### Shape is shape parameter of the four parametric distributions; S0 is  #####
### the survival probability at x; ta and tf are accrual and follow-up     #####
### duration; delta is hazard ratio; eta and zeta are the Bayesian         #####
### posterior confidence level for "GO" and "no-Go" criterion; zeta1 is   #####
```

```
### the Bayesian posterior confidence level for stop futility; frac is a  #####
### information fration for interim analysis; emax is the max number of   #####
### events for iteration of the searching algorithm; dist is option for   #####
### set up reference survival dist as one of the four following  dist.    #####
### including Weibull (WB), log-normal(LN), gamma (GM), log-logistic (LG).#####
##############################################################################
B2Size=function(shape,S0,x,ta,tf,a,delta,eta,zeta,frac,xi,emax,dist)
{
    S1=S0^delta
    tau=ta+tf
    if (dist=="WB"){
     f=function(a,b,u){dweibull(u,a,b)}
     scale1=x/(-log(S1))^(1/shape)
    }

    if (dist=="LN"){
     f=function(a,b,u){dlnorm(u,b,a)}
     scale1=log(x)-shape*qnorm(1-S1)
    }

    if (dist=="LG"){
     f=function(a,b,u){(a/b)*(u/b)^(a-1)/(1+(u/b)^a)^2}
     scale1=x/(1/S1-1)^(1/shape)
    }

    if (dist=="GM"){
     s=function(a,b,u){1-pgamma(u,a,b)}
     f=function(a,b,u){dgamma(u,a,b)}
     root1=function(t){s(shape,t,x)-S1}
     scale1=uniroot(root1,c(0,10))$root
    }
    G=function(u){1-punif(u, tf, tau)}
    g=function(u){f(shape,scale1,u)*G(u)}
    p=integrate(g, 0, tau)$value

 b=2*a/(1+delta)
 zeta=eta*(1-xi)
 for (i in 1:emax){
  f=function(k){
   eta-pgamma(1,shape=a+i,rate=b+k)
  }
 k=uniroot(f,c(0, 999))$root
 m=i; k=round(k,2)
 exp=pgamma(delta,shape=a+i,rate=b+k)
 if (exp<=1-zeta) break
 }
 m1=ceiling(m*frac)
 zeta1=eta*(1-xi*frac)
 f1=function(k){
   zeta1-(1-pgamma(delta,shape=a+m1,rate=b+k))
 }
 k1=uniroot(f1,c(0, 999))$root
 n1=ceiling(m1/p)
 n=ceiling(m/p)
 ans=round(c(a, frac, eta, zeta1, zeta, m1,n1,k1,m,n,k),3)
 return(ans)
}
```

```
B2Size(shape=1,S0=0.53,x=3,ta=4,tf=2,a=2,delta=0.6,
      eta=0.9,frac=0.4,xi=0.01,emax=100,dist="WB")
2  0.4  0.9 0.896 0.891   10   26 10.637   23   59 29.08
```

```
############## Bayesian Two-Stage Design with Frequentist Targets #############
### pow is a function to simulate the frequentitst type I error and power.#####
### optimal is a function to calculate Bayesian two-stage design with the #####
### target frequentitst type I error and power. optimal uses the pow and  #####
### Bayesian two-stage design R code B2Size. In pow function, N is the     #####
### the sample size which is set to be large enough to produce number of   #####
### events more than the required total number of events. M is the total   #####
### number of simulation runs. In optimal function, alphacutoff and        #####
### powercutoff are the cutoffs of frequentist type I error and power.      #####
### other input parameters are same meaning as given in B2Size.            #####
################################################################################
pow=function(S0,x,delta,ta,tf,m1,m2,k1,k2,N,M)
{
  lambda0=-log(S0)/x
  lambda=delta*lambda0
  s0=function(t){exp(-lambda0*t)}
  tau=ta+tf
  set.seed(8232)
  s=0
  for (j in 1:M){
    time=rexp(N, rate=lambda)
    A=runif(N, 0, ta)
    data=data.frame(Entry=A, time=time)
    data0=data[order(data$Entry),]
    time0=data0$time
    A0=data0$Entry
    cens=as.numeric(time0<tau-A0)
    x=pmin(time0,tau-A0)

    d = cumsum(cens)
    s1 = min(which(d == m1))
    s2 = min(which(d == m2))
    x1=x[1:s1]
    cens1=cens[1:s1]
    w1=-log(s0(x1))
    U1=sum(w1)
    x2=x[1:s2]
    w2=-log(s0(x2))
    U2=sum(w2)
    if (U1>k1 & U2>k2) s=s+1
  }
  pow=round(s/M,3)
  return(pow)
}
B2Size=function(shape,S0,x,ta,tf,a,delta,eta,zeta,frac,xi,emax,dist)
{
    S1=S0^delta
    tau=ta+tf
    if (dist=="WB"){
    f=function(a,b,u){dweibull(u,a,b)}
     scale1=x/(-log(S1))^(1/shape)
    }
```

```
  if (dist=="LN"){
   f=function(a,b,u){dlnorm(u,b,a)}
   scale1=log(x)-shape*qnorm(1-S1)
  }

  if (dist=="LG"){
   f=function(a,b,u){(a/b)*(u/b)^(a-1)/(1+(u/b)^a)^2}
   scale1=x/(1/S1-1)^(1/shape)
  }

  if (dist=="GM"){
   s=function(a,b,u){1-pgamma(u,a,b)}
   f=function(a,b,u){dgamma(u,a,b)}
   root1=function(t){s(shape,t,x)-S1}
   scale1=uniroot(root1,c(0,10))$root
  }
  G=function(u){1-punif(u, tf, tau)}
  g=function(u){f(shape,scale1,u)*G(u)}
  p=integrate(g, 0, tau)$value

b=2*a/(1+delta)
zeta=eta*(1-xi)
for (i in 1:emax){
 f=function(k){
  eta-pgamma(1,shape=a+i,rate=b+k)
 }
k=uniroot(f,c(0, 999))$root
m=i; k=round(k,2)
exp=pgamma(delta,shape=a+i,rate=b+k)
if (exp<=1-zeta) break
}
m1=ceiling(m*frac)
zeta1=eta*(1-xi*frac)
f1=function(k){
  zeta1-(1-pgamma(delta,shape=a+m1,rate=b+k))
}
k1=uniroot(f1,c(0, 999))$root
n1=ceiling(m1/p)
n=ceiling(m/p)
ans=round(c(a, frac, eta, zeta1, zeta, m1,n1,k1,m,n,k),3)
return(ans)
}
optimal <- function(alphacutoff, powercutoff, shape, S0, x, ta, tf,
           a, delta, frac, dist){
  eta <- seq(0.8, 0.95, by = 0.05)
  xi <- seq(0.01, 0.15, by = 0.01)
  oc_mat <- NULL
  for (i in 1:length(eta)){
   for (j in 1:length(xi)){
    bsize <- B2Size(shape=shape,S0=S0,x=x,ta=ta,tf=tf,a = a,delta=delta,
             eta=eta[i],frac=frac,xi=xi[j],emax=100,dist = dist)
    ## type I error
    typeI <- pow(S0=S0,x=x,delta=1,ta=ta,tf=tf,m1=bsize[6],m2=bsize[9],
             k1=bsize[8],k2=bsize[11],N=500,M=10000)
    ## power
    power<-pow(S0=S0,x=x,delta=delta,ta=ta,tf=tf,m1=bsize[6],m2=bsize[9],
             k1=bsize[8],k2=bsize[11],N=500,M=10000)
```

```
    if((0.9 * alphacutoff <= typeI) & (typeI <= alphacutoff) &
       (powercutoff <= power) & (power <= powercutoff * 1.025)){
    oc_mat <- rbind(oc_mat, c(eta[i], xi[j], bsize[10], typeI, power))
      }
    }
  }
  out <- oc_mat[oc_mat[, 3] == min(oc_mat[, 3]), ]
  cat("(eta, xi, n) = ", out[1:3], "(alpha, power) = ", out[4:5], "\n")
  return(out)
}
optimal(alphacutoff = 0.05,powercutoff =0.8, shape = 1,S0 = 0.53,x = 3,
  ta = 4, tf = 2, a = 2, delta = 0.6, frac = 0.4, dist = "WB")
## output
(eta, xi, n) =  0.95 0.1 67     (alpha, power) =  0.048 0.803
```

A

Probability of Failure under Uniform Accrual

Assume that subjects are accrued over an accrual period of length t_a, with an additional follow-up time t_f, the study duration $\tau = t_a + t_f$, and the entry time is uniformly distributed over $[0, t_a]$, with no patient loss to follow-up or drop out, Then, the censoring distribution $G(t)$ is a uniform distribution over the interval $[t_f, t_a + t_f]$, that is, $G(t) = 1$ if $t \leq t_f$; $= (t_a + t_f - t)/t_a$ if $t_f \leq t \leq t_a + t_f$; $= 0$ otherwise. Note that $S(t) = \exp\{-\Lambda(t)\}$; and by performing integration calculations, we obtain

$$
\begin{aligned}
p &= \int_0^\infty G(t)S(t)d\Lambda(t) \\
&= \int_0^{t_f} S(t)d\Lambda(t) + \frac{1}{t_a}\int_{t_f}^{\tau}(\tau - t)S(t)d\Lambda(t) \\
&= -\int_0^{t_f} dS(t) - \frac{1}{t_a}\int_{t_f}^{\tau}(\tau - t)dS(t) \\
&= 1 - S(t_f) - \frac{\tau}{t_a}[S(\tau) - S(t_f)] + \frac{1}{t_a}\int_{t_f}^{\tau} t\,dS(t).
\end{aligned}
$$

By integration by parts,

$$
\int_{t_f}^{\tau} t\,dS(t) = \tau S(\tau) - t_f S(t_f) - \int_{t_f}^{\tau} S(t)dt.
$$

Substituting this into the previous equation, we obtain

$$
p = 1 - \frac{1}{t_a}\int_{t_f}^{t_a+t_f} S(t)dt = \frac{1}{t_a}\int_{t_f}^{t_a+t_f} F(t)dt
$$

where $F(t) = 1 - S(t)$ is the cumulative distribution function.

B

Asymptotic Distribution of Nelson-Aalen Estimate of the Cumulative Hazard

Suppose during the accrual phase of the trial, n subjects are enrolled in the study. Let T_i and C_i denote, respectively, the failure time and censoring time of the ith subject, with both being measured from the time of study entry of this subject. We assume that the failure time T_i is independent of the censoring time C_i, and $\{(T_i, C_i); i = 1, \cdots, n\}$ are independent and identically distributed. When the data are examined at the end of the study, we observe the time to failure $X_i = T_i \wedge C_i$ and failure indicator $\Delta_i = I(T_i \leq C_i), i = 1, \cdots, n$. Further define failure process $N_i(t) = \Delta_i I(X_i \leq t)$ and at-risk process $Y_i(t) = I(X_i \geq t)$, and $\bar{N}(t) = \sum_{i=1}^{n} N_i(t)$ and $\bar{Y}(t) = \sum_{i=1}^{n} Y_i(t)$, then, the Nelson-Aalen estimate of the cumulative hazard function is given as follows:

$$\hat{\Lambda}(t) = \int_0^t \frac{d\bar{N}(u)}{\bar{Y}(u)} = \int_0^t \frac{I(\bar{Y}(u) > 0)}{\bar{Y}(u)} d\bar{N}(u)$$

To study the asymptotic distribution of the Nelson-Aalen estimate $\hat{\Lambda}(t)$, let $\Lambda^*(t) = \int_0^t I(\bar{Y}(u) > 0) d\Lambda(u)$, then

$$
\begin{aligned}
\hat{\Lambda}(t) - \Lambda^*(t) &= \int_0^t \frac{I(\bar{Y}(u) > 0)}{\bar{Y}(u)} \{d\bar{N}(u) - \bar{Y}(u) d\Lambda(u)\} \\
&= \sum_{j=1}^{n} \int_0^t \frac{I(\bar{Y}(u) > 0)}{\bar{Y}(u)} dM_i(u) \\
&= \sum_{j=1}^{n} \int_0^t H(u) dM_i(u)
\end{aligned}
$$

where $H(u) = \frac{I(\bar{Y}(u) > 0)}{\bar{Y}(u)}$ is a bounded function and

$$M_i(t) = \sum_{j=1}^{n} \int_0^t \{dN_i(u) - \bar{Y}_i(u) d\Lambda(u)\}$$

is a martingale. Thus, from martingale properties we have $E\{\hat{\Lambda}(t)\} = E\{\Lambda^*(t)\}$ and

$$
\begin{aligned}
\sigma_*^2(t) &= E[\sqrt{n}\{\hat{\Lambda}(t) - \Lambda^*(t)\}]^2 \\
&= nE\left\{\sum_{i=1}^{n}\int_0^t H(u)dM_i(u)\right\}^2 \\
&= nE\left\{\sum_{i=1}^{n}\int_0^t H^2(u)Y_i(u)d\Lambda(u)\right\} \\
&= E\left\{\int_0^t \frac{I(\bar{Y}(u) > 0)}{\bar{Y}(u)/n}d\Lambda(u)\right\}.
\end{aligned}
$$

Let

$$
\hat{\sigma}^2(t) = \int_0^t \frac{I(\bar{Y}(u) > 0)}{\bar{Y}(u)/n}d\hat{\Lambda}(u) = \int_0^t \frac{d\bar{N}(u)}{\bar{Y}^2(u)/n},
$$

it can be shown that $E[\hat{\sigma}^2(t)] = \sigma_*^2(t)$. Thus, $\hat{\sigma}^2(t)$ is a reasonable estimate of $Var\{\sqrt{n}\hat{\Lambda}(t)\}$. We use the following equation

$$
\sqrt{n}\{\hat{\Lambda}(t) - \Lambda^*(t)\} = \sum_{i=1}^{n}\int_0^t H_i(u)dM_i(u)
$$

where $H_i(u) = \sqrt{n}I(\bar{Y}(u) > 0)/\bar{Y}(u)$, and assuming $\pi(t) > 0$. Then, it is easy to verify the following conditions

$$
\begin{aligned}
\sum_{i=1}^{n}\int_0^t H_i^2(u)Y_i(u)d\Lambda(u) &= \sum_{i=1}^{n}\int_0^t \frac{I(\bar{Y}(u) > 0)}{\bar{Y}^2(u)/n}Y_i(u)d\Lambda(u) \\
&= \int_0^t \frac{I(\bar{Y}(u) > 0)}{\bar{Y}(u)/n}d\Lambda(u) \xrightarrow{P} \int_0^t \frac{d\Lambda(u)}{\pi(u)}
\end{aligned}
$$

and

$$
\sum_{i=1}^{n}\int_0^t H_i^2(u)I(|H_i(u)| > \epsilon)Y_i(u)d\Lambda(u) \xrightarrow{P} 0
$$

Furthermore

$$
\sqrt{n}\{\hat{\Lambda}(t) - \Lambda(t)\} = \sqrt{n}\{\hat{\Lambda}(t) - \Lambda^*(t)\} - \sqrt{n}\{\Lambda(t) - \Lambda^*(t)\}
$$

and

$$
\begin{aligned}
\sqrt{n}\{\Lambda(t) - \Lambda^*(t)\} &= \int_0^t \sqrt{n}\{1 - I(\bar{Y}(u) > 0)\}d\Lambda(u) \\
&= \int_0^t \sqrt{n}\{I(\bar{Y}(u) = 0)\}d\Lambda(u)
\end{aligned}
$$

by note $P(\sqrt{n}\{I(\bar{Y}(u) = 0)\} > \epsilon) = P(Y_i(t) = 0, i = 1,\ldots,n) = (1 - \pi(t))^n$, it shows $\sqrt{n}I(\bar{Y}(u) = 0) \xrightarrow{P} 0$, hence $\sqrt{n}\{\Lambda^*(t) - \Lambda(t)\} \xrightarrow{P} 0$, Thus, followed by Slutsky's theorem and Martingale CLT, we have shown that the process $\sqrt{n}\{\hat{\Lambda}(t)-\Lambda(t)\}(0 \leq t \leq \tau)$ converges weakly to a mean-zero Gaussian martingale with variance function

$$\sigma^2(t) = \int_0^t \frac{d\Lambda(u)}{\pi(u)}$$

which can be estimated by

$$\hat{\sigma}^2(t) = \int_0^t \frac{d\bar{N}(u)}{\bar{Y}^2(u)/n}$$

where $\pi(t) = P(X \geq t)$.

C

Derivation Asymptotic Distribution of the OSLRT

First, we calculate the mean and variance of W under the null hypothesis H_0 by noting that $E_{H_0}(O) = nE_{H_0}(\Delta)$ and $E_{H_0}(E) = nE_{H_0}(\Lambda_0(X))$.

Let $f_0(x)$, $S_0(x)$, and $\Lambda_0(x)$ be the density, survival, and cumulative hazard functions of failure time T under the null and $g(x)$ and $G(x)$ be the density and survival functions of censoring time C. Then, by exchange of integrations, we have

$$
\begin{aligned}
E_{H_0}(\Delta) &= \int_0^\infty \left(\int_0^y f_0(x)g(y)dx \right) dy \\
&= \int_0^\infty \left(f_0(x) \int_x^\infty g(y)dy \right) dx \\
&= \int_0^\infty G(x)S_0(x)d\Lambda_0(x).
\end{aligned}
$$

Let $S_X(x)$ be the survival distribution of $X = T \wedge C$ under the null, then $S_X(x) = S_0(x)G(x)$ and by integration by parts, we have

$$
\begin{aligned}
E_{H_0}(\Lambda_0(X)) &= -\int_0^\infty \Lambda_0(x)dS_X(x) \\
&= \int_0^\infty S_X(x)d\Lambda_0(x) \\
&= \int_0^\infty G(x)S_0(x)d\Lambda_0(x).
\end{aligned}
$$

Therefore, the mean of U under the null is $E_{H_0}(U) = \sqrt{n}\{E_{H_0}(\Delta) - E_{H_0}(\Lambda_0(X))\} = 0$. By similar calculation, we have

$$
\begin{aligned}
E_{H_0}(\Delta\Lambda_0(X)) &= \int_0^\infty \left(\int_0^y f_0(x)g(y)\Lambda_0(x)dx \right) dy \\
&= \int_0^\infty \left(f_0(x)\Lambda_0(x) \int_x^\infty g(y)dy \right) dx \\
&= \int_0^\infty G(x)S_0(x)\Lambda_0(x)d\Lambda_0(x),
\end{aligned}
$$

and

$$E_{H_0}(\Lambda_0^2(X)) = -\int_0^\infty \Lambda_0^2(x)dS_X(x)$$

$$= 2\int_0^\infty S_X(x)\Lambda_0(x)d\Lambda_0(x)$$

$$= 2\int_0^\infty G(x)S_0(x)\Lambda_0(x)d\Lambda_0(x).$$

So far, we have shown that $E_{H_0}(\Delta) = E_{H_0}(\Lambda_0(X))$ and $E_{H_0}(\Lambda_0^2(X)) = 2E_{H_0}(\Delta\Lambda_0(X))$. Therefore,

$$\text{Var}_{H_0}(U) = E_{H_0}(\Delta - \Lambda_0(X))^2$$

$$= E_{H_0}(\Delta) - 2E_{H_0}(\Delta\Lambda_0(X)) + E_{H_0}(\Lambda_0^2(X))$$

$$= E_{H_0}(\Delta) = \int_0^\infty G(x)S_0(x)d\Lambda_0(x).$$

Thus,

$$\hat{\sigma}^2 = n^{-1}\sum_{i=1}^n \int_0^\infty Y_i(x)d\Lambda_0(x)$$

is a consistent estimate of $\text{Var}_{H_0}(U)$ under the null and $U/\hat{\sigma} \to N(0,1)$.

Now we derive the exact variance of W under the alternative. Let $f_1(x)$, $S_1(x)$, and $\Lambda_1(x)$ be the density, survival and cumulative hazard functions of failure time T under the alternative. Then by similar calculation, we have

$$E_{H_1}(\Delta) = \int_0^\infty \left(\int_0^y f_1(x)g(y)dx\right)dy$$

$$= \int_0^\infty \left(f_1(x)\int_x^\infty g(y)dy\right)dx$$

$$= \int_0^\infty G(x)S_1(x)d\Lambda_1(x) = p_1(=\sigma_1^2).$$

Let $S_X(x)$ be the survival distribution of $X = T \wedge C$ under the alternative, then $S_X(x) = G(x)S_1(x)$, and by integration by parts, we have

$$E_{H_1}(\Lambda_0(X)) = -\int_0^\infty \Lambda_0(x)dS_X(x)$$

$$= \int_0^\infty S_X(x)d\Lambda_0(x)$$

$$= \int_0^\infty G(x)S_1(x)d\Lambda_0(x) = p_0(=\sigma_0^2).$$

Thus, $E_{H_1}(W) = \sqrt{n}\{E_{H_1}(\Delta) - E_{H_1}(\Lambda_0(X))\} = \sqrt{n}(\sigma_1^2 - \sigma_0^2)$. Similarly, we have

$$
\begin{aligned}
E_{H_1}(\Delta\Lambda_0(X)) &= \int_0^\infty \left(\int_0^y f_1(x)g(y)\Lambda_0(x)dx \right) dy \\
&= \int_0^\infty \left(f_1(x)\Lambda_0(x) \int_x^\infty g(y)dy \right) dx \\
&= \int_0^\infty G(x)S_1(x)\Lambda_0(x)d\Lambda_1(x) = p_{01},
\end{aligned}
$$

and

$$
\begin{aligned}
E_{H_1}(\Lambda_0^2(X)) &= -\int_0^\infty \Lambda_0^2(x)dS_X(x) \\
&= 2\int_0^\infty S_X(x)\Lambda_0(x)d\Lambda_0(x) \\
&= 2\int_0^\infty G(x)S_1(x)\Lambda_0(x)d\Lambda_0(x) = 2p_{00}.
\end{aligned}
$$

Therefore, the exact variance of W under the alternative is given by

$$
\begin{aligned}
\mathrm{Var}_{H_1}(U) &= \mathrm{Var}_{H_1}(\Delta) + \mathrm{Var}_{H_1}(\Lambda_0(X)) - 2\mathrm{Cov}_{H_1}(\Delta, \Lambda_0(X)) \\
&= E_{H_1}(\Delta) - E_{H_1}^2(\delta) + E_{H_1}(\Lambda_0^2(X)) - E_{H_1}^2(\Lambda_0(X)) \\
&\quad - 2E_{H_1}(\Delta\Lambda_0(X)) + 2E_{H_1}(\Delta)E_{H_1}(\Lambda_0(X)) \\
&= p_1 - p_1^2 + 2p_{00} - p_0^2 - 2p_{01} + 2p_0 p_1 = \sigma^2.
\end{aligned}
$$

Since under the alternative H_1, $\hat\sigma^2$ is an consistent estimate of σ_0^2, thus $U - \sqrt{n}\omega \to N(0, \sigma^2)$ and

$$
\frac{U}{\hat\sigma} - \frac{\sqrt{n}\omega}{\sigma_0^2} \to N(0, \frac{\sigma^2}{\sigma_0^2}) \quad \text{as} \quad n \to \infty.
$$

D

Derivation of Equations (6.8) and (6.9)

If (X, Y) is a bivariate normal random vector with mean $\mu = (\mu_1, \mu_2)$ and variance matrix

$$\Sigma = \begin{pmatrix} \sigma_1^2 & \rho\sigma_1\sigma_2 \\ \rho\sigma_1\sigma_2 & \sigma_2^2 \end{pmatrix}$$

then, the conditional distribution of X given $Y = y$ is normal with mean $\mu_1 + (\rho\sigma_1/\sigma_2)(y - \mu_2)$ and variance $\sigma_1^2(1 - \rho^2)$.

As under the null hypothesis H_0, (Z_1, Z) is approximately bivariate normal distributed with mean $\mu = (0, 0)'$ and variance matrix

$$\Sigma = \begin{pmatrix} 1 & \rho_0 \\ \rho_0 & 1 \end{pmatrix}$$

Thus, by conditional distribution integration, we obtain

$$
\begin{aligned}
\alpha &= P(Z_1 > c_1, Z > c \,|H_0) \\
&= \int_{-\infty}^{\infty} \phi(z) P(Z_1 > c_1, Z > c \,|Z = z, H_0) dz \\
&= \int_{c}^{\infty} \phi(z) P(Z_1 > c_1 | Z = z, H_0) dz \\
&= \int_{c}^{\infty} \phi(z) \Phi\left(\frac{\rho_0 z - c_1}{\sqrt{1 - \rho_0^2}} \right) dz.
\end{aligned}
$$

Similarly, under the alternative H_1, (Z_1, Z) is approximately bivariate normal distributed with mean $\mu = (\sqrt{n_1}\omega_1/\sigma_{01}, \sqrt{n}\omega/\sigma_0)'$ and variance matrix

$$\Sigma = \begin{pmatrix} \sigma_{11}^2/\sigma_{01}^2 & \rho_1\sigma_{11}\sigma_1/\sigma_{01}\sigma_0 \\ \rho_1\sigma_{11}\sigma_1/\sigma_{01}\sigma_0 & \sigma_{21}^2/\sigma_0^2 \end{pmatrix}$$

Thus, after standardization, we have

$$\text{power} = P(\bar{Z}_1 > \bar{c}_1, \bar{Z} > \bar{c} \,|H_1)$$

where $\bar{Z}_1 = \frac{\sigma_{01}}{\sigma_{11}}(Z_1 - \frac{\sqrt{n_1}\omega_1}{\sigma_{01}})$ and $\bar{c}_1 = \frac{\sigma_{01}}{\sigma_{11}}(c_1 - \frac{\sqrt{n_1}\omega_1}{\sigma_{01}})$ and $\bar{Z} = \frac{\sigma_0}{\sigma_{21}}(Z - \frac{\sqrt{n}\omega}{\sigma_0})$ and $\bar{c} = \frac{\sigma_0}{\sigma_{21}}(c - \frac{\sqrt{n}\omega}{\sigma_0})$. Thus, $(\bar{Z}_1, \bar{Z})$ is bivariate normal with zero means, unit variances and correlation ρ_1. Again by conditional distribution integration, we

223

have

$$
\begin{aligned}
\text{power} \;&=\; P(\bar{Z}_1 > \bar{c}_1, \bar{Z} > \bar{c} \mid H_1) \\
&=\; \int_{-\infty}^{\infty} \phi(z) P(\bar{Z}_1 > \bar{c}_1, \bar{Z} > \bar{c} \mid \bar{Z} = z, H_1)\, dz \\
&=\; \int_{c}^{\infty} \phi(z) P(\bar{Z}_1 > \bar{c}_1 \mid \bar{Z} = z, H_0)\, dz \\
&=\; \int_{\bar{c}}^{\infty} \phi(z) \Phi\left(\frac{\rho_1 z - \bar{c}_1}{\sqrt{1 - \rho_1^2}}\right) dz.
\end{aligned}
$$

E

Crossing Point for the Mixture Cure Model

Assuming $0 < \pi_0 \leq \pi_1$, hazard ratio $\delta < 1$ and survival distribution $S_0(t)$ is differentiable with density function $f_0(t)$ and both $S_0(t)$ and $f_0(t)$ are positive on interval $[0, \infty)$. The difference of two hazard functions between treatment groups can be expressed as follows:

$$\lambda_0^*(t) - \lambda_1^*(t) = \frac{\lambda_0(t)S_0(t)g(t)}{[\pi_0 + (1 - \pi_0)S_0(t)][\pi_1 + (1 - \pi_1)[S_0(t)]^\delta]}$$

where $g(t) = \pi_1(1 - \pi_0) + (1 - \pi_0)(1 - \pi_1)(1 - \delta)[S_0(t)]^\delta - \pi_0(1 - \pi_1)\delta[S_0(t)]^{\delta - 1}$. The first derivative of $g(t)$ is given by

$$g^{'}(t) = -\delta(1 - \delta)(1 - \pi_1)f_0(t)[S_0(t)]^{\delta - 2}[\pi_0 + (1 - \pi_0)S_0(t)] < 0$$

Thus, $g(t)$ is a monotone decreasing function on $[0, \infty)$. Furthermore it is easy to verify $g(0) = (1 - \pi_0) - (1 - \pi_1)\delta > 0$ and as t goes large/infinity, $[S_0(t)]^\delta$ goes to small/zero and $[S_0(t)]^{\delta - 1}$ goes to large/infinity, therefore, $g(t)$ will be less than zero when t goes to large. Thus, equation $g(t) = 0$ or $\lambda_0^*(t) - \lambda_1^*(t) = 0$ has an unique root on $[0, \infty)$. Therefore, we have shown that the hazard functions between treatment groups has a unique crossing time point.

When $0 \leq \pi_0 < \pi_1$ and $\delta = 1$, the difference of two hazard functions between treatment groups can be expressed as follows:

$$\lambda_0^*(t) - \lambda_1^*(t) = \frac{\lambda_0(t)S_0(t)(\pi_1 - \pi_0)}{[\pi_0 + (1 - \pi_0)S_0(t)][\pi_1 + (1 - \pi_1)S_0(t)]} > 0$$

Thus, there is no crossing over of the hazard functions between two groups.

F

Derivation Asymptotic Distribution of the Score Test Q

By the definition of the test statistic, we have

$$Q = \frac{(\sum_{i=1}^{\tilde{n}} r_i)^2}{\sum_{i=1}^{\tilde{n}} r_i^2} = \frac{(\tilde{n}_2 - \tilde{n}_1)^2}{\tilde{n}},$$

where $\tilde{n}_1$ and $\tilde{n}_2$ are $\#\{i; r_i = -1\}$ and $\#\{i; r_i = 1\}$, respectively, and $\tilde{n} = \tilde{n}_1 + \tilde{n}_2$. By some calculations, we can show an alternative representation of Q as follows:

$$Q = \sum_{k=1}^{2} \frac{\tilde{n}(\hat{p}_k - p_k^0)^2}{p_k^0},$$

where $\hat{p}_k = \tilde{n}_k / \tilde{n}$ and $p_k^0 = P(T_2 > T_1 | H_0) = 0.5, k = 1, 2$. The null hypothesis $H_0 : \gamma = 0$ is equivalent to $H_0 : p_k = p_k^0, k = 1, 2$, where $p_2 = p = P(T_2 > T_1)$ and $p_1 = 1 - p$. Thus, Q is a goodness fit test statistic. By Pearson's theorem, Q converges in distribution to a central chi-square distribution with degree freedom of 1, that is $Q \sim \chi_1^2$.

To calculate sample size, we have to derive the distribution of Q under the alternative. By assuming local alternatives, that is $p_k = p_k^0 + c_k / \sqrt{\tilde{n}}, k = 1, 2$, where c_k is fixed constant as $\tilde{n}$ increases, thus, we have

$$\frac{\sqrt{\tilde{n}}(\hat{p}_k - p_k^0)}{\sqrt{p_k^0}} = \frac{\sqrt{\tilde{n}}(\hat{p}_k - p_k)}{\sqrt{p_k}} \sqrt{\frac{p_k}{p_k^0}} + \frac{\sqrt{\tilde{n}}(p_k - p_k^0)}{\sqrt{p_k^0}}.$$

From the definition of local alternatives, it is easy to see,

$$\sqrt{\frac{p_k}{p_k^0}} = \sqrt{1 + \frac{p_k - p_k^0}{p_k^0}} = \sqrt{1 + \frac{c_k}{p_k^0 \sqrt{\tilde{n}}}} \to 1, \quad \tilde{n} \to \infty$$

and

$$\frac{\sqrt{\tilde{n}}(p_k - p_k^0)}{\sqrt{p_k^0}} = \frac{c_k}{\sqrt{p_k^0}}.$$

Thus, under local alternatives, by the central limit theorem,

$$Z_k = \frac{\sqrt{\tilde{n}}(\hat{p}_k - p_k)}{\sqrt{p_k}} \sqrt{\frac{p_k}{p_k^0}} \sim N(0, 1),$$

where the Z_k are subject to the linear constrain $\sum_{k=1}^{2} Z_k \sqrt{p_k^0} = 0$. Now, we have

$$Q = \sum_{i=1}^{k} (Z_k + \frac{c_k}{\sqrt{p_k^0}})^2.$$

Thus, Q follows a non-central chi-square distribution with degree freedom of 1 and non-central parameter

$$\delta = \sum_{k=1}^{2} \frac{c_k^2}{p_k^0} = \tilde{n} \sum_{k=1}^{2} \frac{(p_k - p_k^0)^2}{p_k^0} = 4\tilde{n}(p - 0.5)^2,$$

where $p = P(T_2 > T_1 | H_a)$.

G

Generate Random Variables from GBVE Model

Let U, V_{11}, V_{12} and M_ν be independent random variables such that $U \sim$ uniform$(0, 1)$, $V_{1i} \sim \exp(1), i = 1, 2$, and $M_\nu = 0$ or 1 with probability $1 - \nu$ and ν, respectively, where $0 < \nu \le 1$. Let $\theta_i, i = 1, 2$ be positive constants. Define $V = V_{11} + M_\nu V_{12}$, $T_1 = \theta_1 U^\nu V$ and $T_2 = \theta_2 (1 - U)^\nu V$, then, as shown by Lee (1979) $(T_1, T_2) \sim \text{GBVE}(\theta_1, \theta_2, \nu)$.

H

Derivation Censoring Survival Distribution of GMI

Assume that patients on current therapy were uniformly enrolled in the study on interval $[0, t_a]$, followed for a period of t_f and no loss to follow-up, then the administrative censoring time C_2 for patients on current therapy is uniformly distributed on interval $[t_f, \tau]$, where $\tau = t_a + t_f$ is the total study duration. We further assume that TTP of the previous therapy T_1 follows the Weibull distribution $S_1(t) = e^{-\lambda t^\kappa}$, which can be estimated from previous therapy TTP data. Then, the survival distribution of GMI censoring time C_2/T_1 can be derived as follows:

$$
\begin{aligned}
G_{\text{GMI}}(u) &= P(C/T_1 > u) \\
&= \int_{t_f}^{\tau} P(C > uT_1 | C = u) f_C(t) dt \\
&= \int_{t_f}^{\tau} P(T_1 < t/u) f_C(t) dt \\
&= 1 - \frac{1}{t_a} \int_{t_f}^{\tau} e^{-\frac{\lambda}{u^\kappa} t^\kappa} dt
\end{aligned}
$$

I

Proof of Monotonicity of the Posterior Probability

Let $X_i \sim f_i = f(x; a, b_i)$ be the gamma density with shape a and rate $b_i, i = 1, 2$, where $b_1 < b_2$. It can be shown that $f_2 - f_1 > 0$ if and only if $x \leq M$, where $M = a(\log b_2 - \log b_1)/(b_2 - b_1) > 0$. We can show that for any constant $C > 0$, $P(X_2 \leq C) > P(X_1 \leq C)$. As

$$P(X_2 \leq C) - P(X_1 \leq C) = \int_0^C (f_2 - f_1) dx.$$

If $C \leq M$, then

$$\int_0^C (f_2 - f_1) dx > 0.$$

If $C > M$, then

$$\int_0^C (f_2 - f_1) dx = \int_0^\infty (f_2 - f_1) dx - \int_C^\infty (f_2 - f_1) dx = -\int_C^\infty (f_2 - f_1) dx > 0.$$

Hence, $P(X_2 \leq C) > P(X_1 \leq C)$ for any $C > 0$.

J

Relationship between Frequentist and Bayesian Type I Error Rates

The maximum likelihood estimate of δ is given by

$$\hat{\delta} = \frac{d}{U}.$$

and its variance estimate can be obtained from the inverse of Fisher information $j^{-1}(\hat{\delta}) = -\{\ell''(\hat{\delta})\}^{-1} = d/U^2$. The Wald test statistic $\hat{\delta}$ is given by

$$Z = \frac{\hat{\delta} - 1}{\sqrt{\hat{Var}(\hat{\delta})}} = \frac{d - U}{\sqrt{d}},$$

which asymptotically has a standard normal distribution and rejects the null hypothesis $Z < -z_{1-\alpha}$. Thus, given the number of events $d = m$, the frequentist type I error rate can be calculated as follows:

$$P(Z < -z_{1-\alpha}|H_0) = P(U > m + \sqrt{m}z_{1-\alpha}|H_0).$$

Then, we calculate the Bayesian posterior probability

$$\inf_{U \geq k} P(\delta < 1|m, U) = P(\delta < 1|m, k),$$

with $k = m + \sqrt{m}z_{1-\alpha}$ and solve the following equation to obtain the frequentist type I error rate α for a Bayesian design with the number of events $d = m$ and posterior probability threshold η,

$$P(\delta < 1|m, m + \sqrt{m}z_{1-\alpha}) = \eta. \tag{J.1}$$

Under the alternative hypothesis $\delta = \delta_1$, given the number of events $d = m$, the frequentist power can be calculated as follows:

$$
\begin{aligned}
1 - \beta &= P(Z < -z_{1-\alpha}|H_1) \\
&\simeq \Phi(k(1 - \delta_1)/\sqrt{m} - z_{1-\alpha}).
\end{aligned}
$$

Thus, we have

$$k = \frac{\sqrt{m}(z_{1-\alpha} + z_{1-\beta})}{1 - \delta_1}.$$

We then calculate the Bayesian posterior probability

$$\inf_{U<k} P(\delta > \delta_1|m, U) = P(\delta > \delta_1|m, k),$$

with $k = \sqrt{m}(z_{1-\alpha}+z_{1-\beta})/(1-\delta_1)$ and solve the following equation to obtain frequentist type II error rate β or power $1-\beta$ for a Bayesian design with the number of events $d = m$ and posterior probability threshold ζ,

$$P(\delta > \delta_1|m, \sqrt{m}(z_{1-\alpha} + z_{1-\beta})/(1 - \delta_1)) = \zeta. \tag{J.2}$$

For example, for the study design with $a = 2$ (shape parameter of the Gamma prior), $\delta_1 = 0.6$, $m = 16$, $\eta = 0.9$ and $\zeta = 0.8$ (first case in Table 9.1), using equation (J.1), the calculated frequentist type I error is $\alpha = 0.109$; and using equation (J.2) with $\alpha = 0.109$, the calculated frequentist power is $1 - \beta = 0.795$.

K

R Code

K.1   Two-Stage Design Using Arcsin-Square Root Transformed Test

```
########### TwoStage.AS using arcsin-square root transformed test ###########
### shape is the shape parameter of the underlying survival distribution;    ###
### S0 and S1 are survival probability at fixed time point x; dist specifies###
### the underlying survival distribution; r is accrual rate per unit time;   ###
### alpha and beta are the type I and type II errors and power=1-beta.        ###
##############################################################################
library(survival)
library(mvtnorm)
TwoStage.AS=function(r, alpha, beta, shape, S0, S1, x, n, dist)
{
   n0 <- FixDes(r,alpha, beta, shape, S0, S1, x)
   n.int=1
   nsearch=seq(n0, n0+20, n.int)
   nvec <- length(nsearch)
   nrho.res <- matrix(rep(NA, nvec * 2), nrow = nvec, ncol = 2)
   nrho.res[, 1] <- nsearch
   n.res <- nsearch
   nvec <- length(n.res)
   min.res <- rep(NA, nvec)
   t1.res <-rep(NA, nvec)
   t2.res <-rep(NA, nvec)
   C1.res <-rep(NA, nvec)
   C2.res <-rep(NA, nvec)
   MTSL.res <-rep(NA, nvec)
   MDA.res <-rep(NA, nvec)
   ES.res <-rep(NA, nvec)
   ETSL.res <-rep(NA, nvec)
   PS0.res <-rep(NA, nvec)
   PS1.res <-rep(NA, nvec)
   n1.res <-rep(NA, nvec)
   n2.res <-rep(NA, nvec)
   u.res<-matrix(rep(NA, nvec * 2), nrow = nvec, ncol = 2)
   se.res<-matrix(rep(NA, nvec * 4), nrow = nvec, ncol = 4)

   tol=10^(-6)

   i=0
   for (n in nsearch){
     i=i+1
```

```
        optout=optimize(f=f.DesWrap, c(0,1),tol=tol,r,alpha,beta,shape,
                    S0, S1, x, n, dist)
            min.res[i] <- optout$objective
            nrho.res[i, 2] <- optout$minimum
            cat("n=", nrho.res[i, 1], "optimal rho=", nrho.res[i,2],
                "ES=", min.res[i], "\n")
            flush.console()

  rho=nrho.res[i,2]
  f.out<-f.Des(rho, r, alpha, beta, shape, S0, S1, x, n, dist)

  MDA.res[i]  <- f.out$mda
  MTSL.res[i] <- f.out$mda+x
  se.res[i,]  <- f.out$se
  u.res[i,]   <- f.out$u
  ETSL.res[i] <- f.out$ETSL
  ES.res[i]   <- f.out$ES
  PS0.res[i]<-f.out$PS0
  PS1.res[i]<-f.out$PS1
  n1.res[i]  <- ceiling(f.out$n1)
  n2.res[i]  <- ceiling(f.out$n2)
  t1.res[i]  <- f.out$t1
  t2.res[i]  <- f.out$t2
  C1.res[i]  <- f.out$C1
  C2.res[i]  <- f.out$C2
}
## truncate result vectors at last sample size evaluated
nrho.res <- nrho.res[1:i,]
min.res <-  min.res[1:i]
n.res <-    n.res[1:i]
t1.res <-   t1.res[1:i]
t2.res <-   t2.res[1:i]
C1.res <-   C1.res[1:i]
C2.res <-   C2.res[1:i]
MTSL.res <- MTSL.res[1:i]
MDA.res <-  MDA.res[1:i]
ETSL.res <- ETSL.res[1:i]
ES.res <-   ES.res[1:i]
PS0.res <-  PS0.res[1:i]
PS1.res <-  PS1.res[1:i]
n1.res <-   n1.res[1:i]
n2.res <-   n2.res[1:i]
se.res <-   se.res[1:i, ]
u.res <-    u.res[1:i, ]


## select the optimal n
order.min <- order(min.res)[1]
n.last <- nrho.res[order.min,1]
outcome <- min.res[order.min]
ETSL<-ETSL.res[order.min]
ES <- ES.res[order.min]
PS0 <- PS0.res[order.min]
PS1 <- PS1.res[order.min]
mda <- MDA.res[order.min]
t1.last <- t1.res[order.min]
t2.last <- t2.res[order.min]
C1.last <- C1.res[order.min]
```

```
  C2.last <- C2.res[order.min]
  n1 <- n1.res[order.min]
  n2 <- n2.res[order.min]
  se <- se.res[order.min,]
  u  <- u.res[order.min,]

  res=structure(
    list(test=c(alpha=alpha,beta=beta,shape=shape,S0=S0,S1=S1,x=x,n0=n0),
    result=c(ES=ES,ETSL=ETSL),StopProb=c(StopProb.Null=PS0,StopProb.Alt=PS1),
    n=c(Stage1.n1=n1,FinalMax.n=n.last),
    boundary=c(C1=C1.last,C2=C2.last),
    stageTime=c(t1=t1.last,t2=t2.last,MTSL=mda+x,MDA=mda)))
  return(res)
}

f.DesWrap=function(rho, r, alpha, beta, shape, S0, S1, x, n, dist)
{
  out<-f.Des(rho, r, alpha, beta, shape, S0, S1, x, n, dist)
  return(out$min)
}

f.Des=function(rho, r, alpha, beta, shape, S0, S1, x, n, dist)
{
    mda=n/r                  #2

    if (dist=="WB"){
     s=function(a,b,u){1-pweibull(u,a,b)}
     f=function(a,b,u){dweibull(u,a,b)}
     h=function(a,b,u){f(a,b,u)/s(a,b,u)}
     scale0=x/(-log(S0))^(1/shape)
     scale1=x/(-log(S1))^(1/shape)
     }

    if (dist=="LN"){
     s=function(a,b,u){1-plnorm(u,b,a)}
     f=function(a,b,u){dlnorm(u,b,a)}
     h=function(a,b,u){f(a,b,u)/s(a,b,u)}
     scale0=log(x)-shape*qnorm(1-S0)
     scale1=log(x)-shape*qnorm(1-S1)
     }

    if (dist=="LG"){
     s=function(a,b,u){1/(1+(u/b)^a)}
     f=function(a,b,u){(a/b)*(u/b)^(a-1)/(1+(u/b)^a)^2}
     h=function(a,b,u){f(a,b,u)/s(a,b,u)}
     scale0=x/(1/S0-1)^(1/shape)
     scale1=x/(1/S1-1)^(1/shape)
     }

    if (dist=="GM"){
     s=function(a,b,u){1-pgamma(u,a,b)} ## shape=a; scale=b
     f=function(a,b,u){dgamma(u,a,b)}
     h=function(a,b,u){f(a,b,u)/s(a,b,u)}
     root0=function(t){s(shape,t,x)-S0}
     scale0=uniroot(root0,c(0,10))$root
     root1=function(t){s(shape,t,x)-S1}
     scale1=uniroot(root1,c(0,10))$root
```

```
}

sig21 <- sqrt((1 - S1)/S1)          #3
sig20 <- sqrt((1 - S0)/S0)          #3
sig11=sig21/rho                     #4   rho need update in iteration
f.int=function(u,r,shape,scale,x,n,t,mda){
  h(shape,scale,u)/(s(shape,scale,u)*(1-punif(u, t-mda, t)))
}
fG=function(t,r,shape,scale,x,n,mda,sig11){
   integrate(f.int,0,x,r,shape,scale,x,n,t,mda)$value-sig11^2
}
epsilon=10^(-5)
t1.root=uniroot(fG,c(x+epsilon,mda+x),r,shape,scale1,x,n,mda,sig11)#5
t1=t1.root$root
t2=max(mda-t1,0)                    #6
g.int=function(u,r,shape,scale,x,n,mda,t1){
  h(shape, scale, u)/(s(shape,scale,u)*(1-punif(u, t1-mda, t1)))
}
sig10=sqrt(integrate(g.int,0,x,r,shape,scale0,x,n,mda,t1)$value)#7
rho0=sig20/sig10                    #8
v2=S1*sig21^2/(4*(1-S1))
u2=sqrt(n)*(asin(sqrt(S1))-asin(sqrt(S0)))/sqrt(v2)    #9
u1=rho*u2
u=c(u1,u2)

sigma1=matrix(c(1,rho,rho,1),2,2)
C1low<- -3.5
C1up<-qnorm(1-alpha)
C2.init<- qnorm(1-alpha)
powf=function(C1, C2.init, sigma1,beta,u1,u2)
  {pmvnorm(lower=c((C1-u1),(C2.init-u2)),
           upper=c(Inf,Inf),sigma=sigma1)-(1-beta)}
C1root <- uniroot(powf,c(C1low,C1up),
   C2=C2.init,sigma1=sigma1,beta=beta,u1=u1,u2=u2)
C1.init <- C1root$root
par.init=c(C1.init,C2.init)
f=function(x){
 x1=x[1]
 x2=x[2]
 A=pmvnorm(lower=c(x1,x2),upper=c(Inf, Inf), mean=c(0,0),
           corr=matrix(c(1,rho0,rho0,1),2,2))-alpha
 B=pmvnorm(lower=c(x1,x2),upper=c(Inf, Inf), mean=c(u1,u2),
           corr=matrix(c(1,rho,rho,1),2,2))-(1-beta)
 A^2+B^2
}
fit=optim(par=par.init,f)
C1=fit$par[1]
C2=fit$par[2]
PS0=pnorm(C1)
PS1=pnorm(C1-u1)
n1=t1*r
n2=t2*r
ETSL=t1+(1-PS0)*(mda+x-t1)
ES=n1+(1-PS0)*n2        #11

return(list(min=ES,ETSL=ETSL,ES=ES,PS0=PS0,PS1=PS1,n1=n1,n2=n2,t1=t1,
  t2=t2, mda=mda,C1=C1,C2=C2,se=c(sig10,sig20,sig11,sig21),u=u))
```

```
}
FixDes=function(r,alpha,beta,shape,S0, S1, x){
    sig21 <- sqrt((1 - S1)/S1)
    v2=S1*sig21^2/(4*(1-S1))
    z0=qnorm(1-alpha); z1=qnorm(1-beta)
    n0=(z0+z1)^2*v2/(asin(sqrt(S1))-asin(sqrt(S0)))^2
    return(ceiling(n0))
}
TwoStage.AS(shape=1,S0=0.35,S1=0.5,x=12,r=2,alpha=0.05,beta=0.2,dist="WB")
```

K.2   Two-Stage Design Using MOSLRT with Restricted Follow-up

```
#################### Optimal.rKJ.MOSLRT Input parameters #####################
###   shape is the shape parameter for one the four parametric distributions;###
###   S0 is the survival probability at fixed time point x0 under the null;  ###
###   hr is inverse of hazard ratio; x is fixed follow-up time period;       ###
###   rate is constant accrual rate; alpha and beta are type I and II errors ###
###   dist is distribution option with 'WB' as Weibull, 'GM' as Gamma,'LN'   ###
###   as log-normal, 'LG' as log-logistic.                                   ###
#############################################################################
library(survival)
Optimal.rKJ.MOSLRT<-function(shape,S0,x0,hr,x,rate,alpha,beta,dist)
{
  calculate_alpha<-function(c2, c1, rho0){
    fun1<-function(z, c1, rho0){
      f<-dnorm(z)*pnorm((rho0*z-c1)/sqrt(1-rho0^2))
      return(f)
    }
    alpha<-integrate(fun1, lower= c2, upper= Inf, c1, rho0)$value
    return(alpha)
  }

  calculate_power<-function(cb, cb1, rho1){
    fun2<-function(z, cb1, rho1){
      f<-dnorm(z)*pnorm((rho1*z-cb1)/sqrt(1-rho1^2))
      return(f)
    }
    pwr<-integrate(fun2,lower=cb,upper=Inf,cb1=cb1,rho1=rho1)$value
    return(pwr)
  }

  fct<-function(zi, ceps=0.0001,alphaeps=0.0001,nbmaxiter=100,dist){
    ta<-as.numeric(zi[1])
    t1<-as.numeric(zi[2])
    c1<-as.numeric(zi[3])

  if (dist=="WB"){
    s0=function(u){1-pweibull(u,shape,scale0)}
    f0=function(u){dweibull(u,shape,scale0)}
    h0=function(u){f0(u)/s0(u)}
```

```
      H0=function(u){-log(s0(u))}
      s=function(b,u) {s0(u)^b}
      h=function(b,u){b*f0(u)/s0(u)}
      H=function(b,u){-b*log(s0(u))}
      scale0=x0/(-log(S0))^(1/shape)
      scale1=hr
  }

  if (dist=="LN"){
    s0=function(u){1-plnorm(u,scale0,shape)}
    f0=function(u){dlnorm(u,scale0,shape)}
    h0=function(u){f0(u)/s0(u)}
    H0=function(u){-log(s0(u))}
    s=function(b,u) {s0(u)^b}
    h=function(b,u){b*f0(u)/s0(u)}
    H=function(b,u){-b*log(s0(u))}
    scale0=log(x0)-shape*qnorm(1-S0)
    scale1=hr
  }

  if (dist=="LG"){
    s0=function(u){1/(1+(u/scale0)^shape)}
    f0=function(u){(shape/scale0)*(u/scale0)^(shape-1)/(1+(u/scale0)^shape)^2}
    h0=function(u){f0(u)/s0(u)}
    H0=function(u){-log(s0(u))}
    s=function(b,u) {s0(u)^b}
    h=function(b,u){b*f0(u)/s0(u)}
    H=function(b,u){-b*log(s0(u))}
    scale0=x0/(1/S0-1)^(1/shape)
    scale1=hr
  }

  if (dist=="GM"){
    s0=function(u){1-pgamma(u,shape,scale0)}
    f0=function(u){dgamma(u,shape,scale0)}
    h0=function(u){f0(u)/s0(u)}
    H0=function(u){-log(s0(u))}
    s=function(b,u) {s0(u)^b}
    h=function(b,u){b*f0(u)/s0(u)}
    H=function(b,u){-b*log(s0(u))}
    root0=function(t){1-pgamma(x0,shape,t)-S0}
    scale0=uniroot(root0,c(0,10))$root
    scale1=hr
  }

  g0=function(t){s(scale1,t)*h0(t)}
  g1=function(t){s(scale1,t)*h(scale1,t)}
  g00=function(t){s(scale1,t)*H0(t)*h0(t)}
  g01=function(t){s(scale1,t)*H0(t)*h(scale1,t)}
  p0=integrate(g0, 0, x)$value
  p1=integrate(g1, 0, x)$value
  p00=integrate(g00, 0, x)$value
  p01=integrate(g01, 0, x)$value
  sigma2.1=p1-p1^2+2*p00-p0^2-2*p01+2*p0*p1
  sigma2.0=(p0+p1)/2
  om=p0-p1
```

```
G1=function(t){1-punif(t, t1-ta, t1)}
g0=function(t){s(scale1,t)*h0(t)*G1(t)}
g1=function(t){s(scale1,t)*h(scale1,t)*G1(t)}
g00=function(t){s(scale1,t)*H0(t)*h0(t)*G1(t)}
g01=function(t){s(scale1,t)*H0(t)*h(scale1,t)*G1(t)}
p0=integrate(g0, 0, x)$value
p1=integrate(g1, 0, x)$value
p00=integrate(g00, 0, x)$value
p01=integrate(g01, 0, x)$value
sigma2.11=p1-p1^2+2*p00-p0^2-2*p01+2*p0*p1
sigma2.01=(p0+p1)/2
om1=p0-p1

q1=function(t){s0(t)*h0(t)*G1(t)}
q=function(t){s0(t)*h0(t)}
v1=integrate(q1, 0, x)$value
v=integrate(q, 0, x)$value
rho0=sqrt(v1/v)
rho1<-sqrt(sigma2.11/sigma2.1)

cL<-(-10)
cU<-(10)
alphac<-100
iter<-0
while ((abs(alphac-alpha)>alphaeps|cU-cL>ceps)&iter<nbmaxiter){
  iter<-iter+1
  c<-(cL+cU)/2
  alphac<-calculate_alpha(c, c1, rho0)
  if (alphac>alpha) {
    cL<-c
  } else {
    cU<-c
  }
}

cb1<-sqrt((sigma2.01/sigma2.11))*(c1-(om1*sqrt(rate*t1)/sqrt(sigma2.01)))
cb<-sqrt((sigma2.0/sigma2.1))*(c-(om*sqrt(rate*ta))/sqrt(sigma2.0))
pwrc<-calculate_power(cb=cb, cb1=cb1, rho1=rho1)
res<-c(cL, cU, alphac, 1-pwrc, rho0, rho1, cb1, cb)
return(res)
}

c1<-0 ; rho0<-0; cb1<-0; rho1<-0 ; hz<-c(0,0);
ceps<-0.001;alphaeps<-0.001;nbmaxiter<-100

if (dist=="WB"){
  s0=function(u){1-pweibull(u,shape,scale0)}
  f0=function(u){dweibull(u,shape,scale0)}
  h0=function(u){f0(u)/s0(u)}
  H0=function(u){-log(s0(u))}
  s=function(b,u) {s0(u)^b}
  h=function(b,u){b*f0(u)/s0(u)}
  H=function(b,u){-b*log(s0(u))}
  scale0=x0/(-log(S0))^(1/shape)
  scale1=hr
}
```

```
if (dist=="LN"){
  s0=function(u){1-plnorm(u,scale0,shape)}
  f0=function(u){dlnorm(u,scale0,shape)}
  h0=function(u){f0(u)/s0(u)}
  H0=function(u){-log(s0(u))}
  s=function(b,u) {s0(u)^b}
  h=function(b,u){b*f0(u)/s0(u)}
  H=function(b,u){-b*log(s0(u))}
  scale0=log(x0)-shape*qnorm(1-S0)
  scale1=hr
}

if (dist=="LG"){
  s0=function(u){1/(1+(u/scale0)^shape)}
  f0=function(u){(shape/scale0)*(u/scale0)^(shape-1)/(1+(u/scale0)^shape)^2}
  h0=function(u){f0(u)/s0(u)}
  H0=function(u){-log(s0(u))}
  s=function(b,u) {s0(u)^b}
  h=function(b,u){b*f0(u)/s0(u)}
  H=function(b,u){-b*log(s0(u))}
  scale0=x0/(1/S0-1)^(1/shape)
  scale1=hr
}

if (dist=="GM"){
  s0=function(u){1-pgamma(u,shape,scale0)}
  f0=function(u){dgamma(u,shape,scale0)}
  h0=function(u){f0(u)/s0(u)}
  H0=function(u){-log(s0(u))}
  s=function(b,u) {s0(u)^b}
  h=function(b,u){b*f0(u)/s0(u)}
  H=function(b,u){-b*log(s0(u))}
  root0=function(t){1-pgamma(x0,shape,t)-S0}
  scale0=uniroot(root0,c(0,10))$root
  scale1=hr
}

g0=function(t){s(scale1,t)*h0(t)}
g1=function(t){s(scale1,t)*h(scale1,t)}
g00=function(t){s(scale1,t)*H0(t)*h0(t)}
g01=function(t){s(scale1,t)*H0(t)*h(scale1,t)}
p0=integrate(g0, 0, x)$value
p1=integrate(g1, 0, x)$value
p00=integrate(g00, 0, x)$value
p01=integrate(g01, 0, x)$value
s1=sqrt(p1-p1^2+2*p00-p0^2-2*p01+2*p0*p1)
s0=sqrt((p0+p1)/2)
om=p0-p1
nsingle<-(s0*qnorm(1-alpha)+s1*qnorm(1-beta))^2/om^2
nsingle<-ceiling(nsingle)
tasingle<-nsingle/rate

Single_stage<-data.frame(n.single=nsingle,ta.single=tasingle,
                         c.single=qnorm(1-alpha))
atc0<-data.frame(n=nsingle, t1=tasingle, c1=0.25)
nbpt<-11
pascote<-1.26
```

```
cote<-1*pascote
c1.lim<-nbpt*c(-1,1)
EnH0<-10000
iter<-0

while (iter<nbmaxiter & diff(c1.lim)/nbpt>0.001){
iter<-iter+1
cat("iter=",iter,"&EnH0=",round(EnH0, 2),"& Dc1/nbpt=",
    round(diff(c1.lim)/nbpt,5),"\n",sep="")

if (iter%%2==0) nbpt<-nbpt+1
cote<-cote/pascote
n.lim<-atc0$n+c(-1, 1)*nsingle*cote
t1.lim<-atc0$t1+c(-1, 1)*tasingle*cote
c1.lim<-atc0$c1+c(-1, 1)*cote
ta.lim<-n.lim/rate
t1.lim<-pmax(0, t1.lim)
n<-seq(n.lim[1], n.lim[2], l=nbpt)
n<-ceiling(n)
n<-unique(n)
ta<-n/rate
t1<-seq(t1.lim[1], t1.lim[2], l=nbpt)
c1<-seq(c1.lim[1], c1.lim[2], l=nbpt)
ta<-ta[ta>0]
t1<-t1[t1>=0.2*tasingle & t1<=1.2*tasingle]
z<-expand.grid(list(ta=ta, t1=t1, c1=c1))
z<-z[z$ta>z$t1,]
nz<-dim(z)[1]
z$pap<-pnorm(z$c1)
z$eta<-z$ta-pmax(0, z$ta-z$t1)*z$pap
z$enh0<-z$eta*rate
z<-z[z$enh0<=EnH0,]
nz1<-dim(z)[1]
resz<-t(apply(z,1,fct,ceps=ceps,alphaeps=alphaeps,nbmaxiter=nbmaxiter,
        dist=dist))
resz<-as.data.frame(resz)
names(resz)<-c("cL","cU","alphac","betac","rho0","rho1","cb1","cb")
r<-cbind(z, resz)
r$pap<-pnorm(r$c1)
r$eta<-r$ta-pmax(0, r$ta-r$t1)*r$pap
r$etar<-r$eta*rate
r$tar<-r$ta*rate
r$c<-r$cL+(r$cU-resz$cL)/2
r$diffc<-r$cU-r$cL
r$diffc<-ifelse(r$diffc<=ceps, 1, 0)
r<-r[1-r$betac>=1-beta,]
r<-r[order(r$enh0),]
r$n<-r$ta*rate

if (dim(r)[1]>0) {
  atc<-r[, c("ta", "t1", "c1", "n")][1,]

  r1<-r[1,]
  if (r1$enh0<EnH0) {
    EnH0<-r1$enh0
    atc0<-atc
  }
```

```
    } else {
      atc<-data.frame(ta=NA, t1=NA, c1=NA, n=NA)
    }
    atc$iter<-iter
    atc$enh0<-r$enh0[1]
    atc$EnH0<-EnH0
    atc$tai<-ta.lim[1]
    atc$tas<-ta.lim[2]
    atc$ti<-t1.lim[1]
    atc$ts<-t1.lim[2]
    atc$ci<-c1.lim[1]
    atc$cs<-c1.lim[2]
    atc$cote<-cote
    if (iter==1) {
      atcs<-atc
    } else {
      atcs<-rbind(atcs, atc)
    }
  }
  atcs$i<-1:dim(atcs)[1]
  a<-atcs[!is.na(atcs$n),]

  p<-t(apply(a,1,fct,ceps=ceps,alphaeps=alphaeps,nbmaxiter=nbmaxiter,
        dist=dist))
  p<-as.data.frame(p)
  names(p)<-c("cL","cU","alphac","betac","rho0","rho1","cb1","cb")
  p$i<-a$i

  res<-merge(atcs, p, by="i", all=T)
  res$pap<-round(pnorm(res$c1),4)  ## stopping prob of stage 1 ##
  res$ta<-res$n/rate
  res$EnH0<-(res$ta-pmax(0, res$ta-res$t1)*res$pap)*rate
  res$diffc<-res$cU-res$cL
  res$c<-round(res$cL+(res$cU-res$cL)/2,4)
  res$diffc<-ifelse(res$diffc<=ceps, 1, 0)
  res<-res[order(res$enh0),]
  des<-round(res[1,],4)

  if (dist=="SP"){
    param<-data.frame(hr=hr, alpha=alpha, beta=beta, rate=rate, x=x)}
   else {
    param<-data.frame(shape=shape,S0=S0,hr=hr,alpha=alpha,beta=beta,
                       rate=rate, x0=x0, x=x) }
  Two_stage<- data.frame(n1= ceiling(des$t1*param$rate), c1=des$c1,
               n=ceiling(des$ta*param$rate), c=des$c, t1=des$t1,
               MTSL=des$ta+param$x, ES=des$EnH0, PS=des$pap)
  DESIGN<-list(param=param,Single_stage=Single_stage,Two_stage=Two_stage)
  return(DESIGN)
}
Optimal.rKJ.MOSLRT(shape=1.47327,S0=0.5,x0=3.5,hr=0.5913,x=5,rate=2,alpha=0.05,
         beta=0.2,dist="WB")
```

K.3 Two-Stage Design Using MOSLRT without Restricted Follow-up

```
################# Optimal.KJ.MOSLRT Input Parameters #########################
###   shape is the shape parameter for one the four parametric distributions;###
###   S0 is the survival probability at fixed time point x0 under the null;  ###
###   hr is inverse of hazard ratio; rate is constant accrual rate; alpha and###
###   beta are type I and II errors; dist is distribution option with 'WB' as###
###   Weibull, 'GM' as Gamma, 'LN' as log-normal, 'LG' as log-logistic.    ###
###############################################################################
library(survival)
Optimal.KJ.MOSLRT<-function(shape,S0,x0,hr,tf,rate,alpha,beta,dist)
{
  calculate_alpha<-function(c2, c1, rho0){
    fun1<-function(z, c1, rho0){
      f<-dnorm(z)*pnorm((rho0*z-c1)/sqrt(1-rho0^2))
      return(f)
    }
    alpha<-integrate(fun1, lower= c2, upper= Inf, c1, rho0)$value
    return(alpha)
  }

  calculate_power<-function(cb, cb1, rho1){
    fun2<-function(z, cb1, rho1){
      f<-dnorm(z)*pnorm((rho1*z-cb1)/sqrt(1-rho1^2))
      return(f)
    }
    pwr<-integrate(fun2,lower=cb,upper=Inf,cb1=cb1,rho1=rho1)$value
    return(pwr)
  }

  fct<-function(zi, ceps=0.0001,alphaeps=0.0001,nbmaxiter=100,dist){
    ta<-as.numeric(zi[1])
    t1<-as.numeric(zi[2])
    c1<-as.numeric(zi[3])

  if (dist=="WB"){
    f0=function(t){(shape/scale0)*(t/scale0)^(shape-1)*exp(-(t/scale0)^shape)}
    s0=function(t){exp(-(t/scale0)^shape)}
    h0=function(t){(shape/scale0)*(t/scale0)^(shape-1)}
    H0=function(t){(t/scale0)^shape}
    s=function(b, t){exp(-b*(t/scale0)^shape)}
    h=function(b,t){b*h0(t)}
    H=function(b,t){b*H0(t)}
    scale0=x0/(-log(S0))^(1/shape)
    scale1=hr
  }


  if (dist=="LN"){
    s0=function(u){1-plnorm(u,scale0,shape)}
    f0=function(u){dlnorm(u,scale0,shape)}
    h0=function(u){f0(u)/s0(u)}
    H0=function(u){-log(s0(u))}
```

```
       s=function(b,u) {s0(u)^b}
       h=function(b,u){b*f0(u)/s0(u)}
       H=function(b,u){-b*log(s0(u))}
       scale0=log(x0)-shape*qnorm(1-S0)
       scale1=hr
}

  if (dist=="LG"){
     s0=function(u){1/(1+(u/scale0)^shape)}
     f0=function(u){(shape/scale0)*(u/scale0)^(shape-1)/(1+(u/scale0)^shape)^2}
     h0=function(u){f0(u)/s0(u)}
     H0=function(u){-log(s0(u))}
     s=function(b,u) {s0(u)^b}
     h=function(b,u){b*f0(u)/s0(u)}
     H=function(b,u){-b*log(s0(u))}
     scale0=x0/(1/S0-1)^(1/shape)
     scale1=hr
  }

  if (dist=="GM"){
     s0=function(u){1-pgamma(u,shape,scale0)}
     f0=function(u){dgamma(u,shape,scale0)}
     h0=function(u){f0(u)/s0(u)}
     H0=function(u){-log(s0(u))}
     s=function(b,u) {s0(u)^b}
     h=function(b,u){b*f0(u)/s0(u)}
     H=function(b,u){-b*log(s0(u))}
     root0=function(t){1-pgamma(x0,shape,t)-S0}
     scale0=uniroot(root0,c(0,10))$root
     scale1=hr
  }

  G=function(t){1-punif(t, tf, ta+tf)}
  g0=function(t){s(scale1,t)*h0(t)*G(t)}
  g1=function(t){s(scale1,t)*h(scale1,t)*G(t)}
  g00=function(t){s(scale1,t)*H0(t)*h0(t)*G(t)}
  g01=function(t){s(scale1,t)*H0(t)*h(scale1,t)*G(t)}
  p0=integrate(g0, 0, ta+tf)$value
  p1=integrate(g1, 0, ta+tf)$value
  p00=integrate(g00, 0, ta+tf)$value
  p01=integrate(g01, 0, ta+tf)$value
  sigma2.1=p1-p1^2+2*p00-p0^2-2*p01+2*p0*p1
  sigma2.0=(p0+p1)/2
  om=p0-p1

  G1=function(t){1-punif(t, t1-ta, t1)}
  g0=function(t){s(scale1,t)*h0(t)*G1(t)}
  g1=function(t){s(scale1,t)*h(scale1,t)*G1(t)}
  g00=function(t){s(scale1,t)*H0(t)*h0(t)*G1(t)}
  g01=function(t){s(scale1,t)*H0(t)*h(scale1,t)*G1(t)}
  p0=integrate(g0, 0, ta+tf)$value
  p1=integrate(g1, 0, ta+tf)$value
  p00=integrate(g00, 0, ta+tf)$value
  p01=integrate(g01, 0, ta+tf)$value
  sigma2.11=p1-p1^2+2*p00-p0^2-2*p01+2*p0*p1
  sigma2.01=(p0+p1)/2
  om1=p0-p1
```

```
q1=function(t){s0(t)*h0(t)*G1(t)}
q=function(t){s0(t)*h0(t)*G(t)}
v1=integrate(q1, 0, ta+tf)$value
v=integrate(q, 0, ta+tf)$value
rho0=sqrt(v1/v)
rho1<-sqrt(sigma2.11/sigma2.1)

cL<-(-10)
cU<-(10)
alphac<-100
iter<-0
while ((abs(alphac-alpha)>alphaeps|cU-cL>ceps)&iter<nbmaxiter){
  iter<-iter+1
  c<-(cL+cU)/2
  alphac<-calculate_alpha(c, c1, rho0)
  if (alphac>alpha) {
    cL<-c
  } else {
    cU<-c
  }
}

cb1<-sqrt((sigma2.01/sigma2.11))*(c1-(om1*sqrt(rate*t1)/sqrt(sigma2.01)))
cb<-sqrt((sigma2.0/sigma2.1))*(c-(om*sqrt(rate*ta))/sqrt(sigma2.0))
pwrc<-calculate_power(cb=cb, cb1=cb1, rho1=rho1)
res<-c(cL, cU, alphac, 1-pwrc, rho0, rho1, cb1, cb)
return(res)
}

c1<-0 ; rho0<-0; cb1<-0; rho1<-0 ; hz<-c(0,0);
ceps<-0.001;alphaeps<-0.001;nbmaxiter<-100

Duration=function(shape,S0,x0,hr,tf,rate,alpha,beta,dist)
{
 if (dist=="WB"){
   f0=function(t){(shape/scale0)*(t/scale0)^(shape-1)*exp(-(t/scale0)^shape)}
   s0=function(t){exp(-(t/scale0)^shape)}
   h0=function(t){(shape/scale0)*(t/scale0)^(shape-1)}
   H0=function(t){(t/scale0)^shape}
   s=function(b, t){exp(-b*(t/scale0)^shape)}
   h=function(b,t){b*h0(t)}
   H=function(b,t){b*H0(t)}
   scale0=x0/(-log(S0))^(1/shape)
   scale1=hr
 }

 if (dist=="LN"){
   s0=function(u){1-plnorm(u,scale0,shape)}
   f0=function(u){dlnorm(u,scale0,shape)}
   h0=function(u){f0(u)/s0(u)}
   H0=function(u){-log(s0(u))}
   s=function(b,u) {s0(u)^b}
   h=function(b,u){b*f0(u)/s0(u)}
   H=function(b,u){-b*log(s0(u))}
   scale0=log(x0)-shape*qnorm(1-S0)
```

```
      scale1=hr
  }

  if (dist=="LG"){
    s0=function(u){1/(1+(u/scale0)^shape)}
    f0=function(u){(shape/scale0)*(u/scale0)^(shape-1)/(1+(u/scale0)^shape)^2}
    h0=function(u){f0(u)/s0(u)}
    H0=function(u){-log(s0(u))}
    s=function(b,u) {s0(u)^b}
    h=function(b,u){b*f0(u)/s0(u)}
    H=function(b,u){-b*log(s0(u))}
    scale0=x0/(1/S0-1)^(1/shape)
    scale1=hr
  }

  if (dist=="GM"){
    s0=function(u){1-pgamma(u,shape,scale0)}
    f0=function(u){dgamma(u,shape,scale0)}
    h0=function(u){f0(u)/s0(u)}
    H0=function(u){-log(s0(u))}
    s=function(b,u) {s0(u)^b}
    h=function(b,u){b*f0(u)/s0(u)}
    H=function(b,u){-b*log(s0(u))}
    root0=function(t){1-pgamma(x0,shape,t)-S0}
    scale0=uniroot(root0,c(0,10))$root
    scale1=hr
  }

    root=function(ta){
      tau=ta+tf
      G=function(t){1-punif(t, tf, tau)}
      g0=function(t){s(scale1,t)*h0(t)*G(t)}
      g1=function(t){s(scale1,t)*h(scale1,t)*G(t)}
      g00=function(t){s(scale1,t)*H0(t)*h0(t)*G(t)}
      g01=function(t){s(scale1,t)*H0(t)*h(scale1,t)*G(t)}
      p0=integrate(g0, 0, tau)$value
      p1=integrate(g1, 0, tau)$value
      p00=integrate(g00, 0, tau)$value
      p01=integrate(g01, 0, tau)$value
      s1=sqrt(p1-p1^2+2*p00-p0^2-2*p01+2*p0*p1)
      s0=sqrt((p0+p1)/2)
      om=p0-p1
      rate*ta-(s0*qnorm(1-alpha)+s1*qnorm(1-beta))^2/om^2
      }
      tasingle=uniroot(root, lower=0, upper=5*tf)$root
      nsingle<-ceiling(tasingle*rate)
      tasingle=ceiling(tasingle)
      ans=list(nsingle=nsingle, tasingle=tasingle)
      return(ans)
  }

  ans=Duration(shape,S0,x0,hr,tf,rate,alpha,beta,dist)
  tasingle<-ans$tasingle
  nsingle=ans$nsingle

  Single_stage<-data.frame(n.single=nsingle,ta.single=tasingle,
                           c.single=qnorm(1-alpha))
```

```
atc0<-data.frame(n=nsingle, t1=tasingle, c1=0.25)
nbpt<-11
pascote<-1.26
cote<-1*pascote
c1.lim<-nbpt*c(-1,1)
EnH0<-10000
iter<-0

while (iter<nbmaxiter & diff(c1.lim)/nbpt>0.001){
iter<-iter+1
cat("iter=",iter,"&EnH0=",round(EnH0, 2),"& Dc1/nbpt=",
    round(diff(c1.lim)/nbpt,5),"\n",sep="")

if (iter%%2==0) nbpt<-nbpt+1
cote<-cote/pascote
n.lim<-atc0$n+c(-1, 1)*nsingle*cote
t1.lim<-atc0$t1+c(-1, 1)*tasingle*cote
c1.lim<-atc0$c1+c(-1, 1)*cote
ta.lim<-n.lim/rate
t1.lim<-pmax(0, t1.lim)
n<-seq(n.lim[1], n.lim[2], l=nbpt)
n<-ceiling(n)
n<-unique(n)
ta<-n/rate
t1<-seq(t1.lim[1], t1.lim[2], l=nbpt)
c1<-seq(c1.lim[1], c1.lim[2], l=nbpt)
ta<-ta[ta>0]
t1<-t1[t1>=0.2*tasingle & t1<=1.2*tasingle]
z<-expand.grid(list(ta=ta, t1=t1, c1=c1))
z<-z[z$ta>z$t1,]
nz<-dim(z)[1]
z$pap<-pnorm(z$c1)
z$eta<-z$ta-pmax(0, z$ta-z$t1)*z$pap
z$enh0<-z$eta*rate
z<-z[z$enh0<=EnH0,]
nz1<-dim(z)[1]
resz<-t(apply(z,1,fct,ceps=ceps,alphaeps=alphaeps,nbmaxiter=nbmaxiter,
        dist=dist))
resz<-as.data.frame(resz)
names(resz)<-c("cL","cU","alphac","betac","rho0","rho1","cb1","cb")
r<-cbind(z, resz)
r$pap<-pnorm(r$c1)
r$eta<-r$ta-pmax(0, r$ta-r$t1)*r$pap
r$etar<-r$eta*rate
r$tar<-r$ta*rate
r$c<-r$cL+(r$cU-resz$cL)/2
r$diffc<-r$cU-r$cL
r$diffc<-ifelse(r$diffc<=ceps, 1, 0)
r<-r[1-r$betac>=1-beta,]
r<-r[order(r$enh0),]
r$n<-r$ta*rate

if (dim(r)[1]>0) {
  atc<-r[, c("ta", "t1", "c1", "n")][1,]

  r1<-r[1,]
  if (r1$enh0<EnH0) {
```

```
          EnHO<-r1$enh0
          atc0<-atc
      }
   } else {
      atc<-data.frame(ta=NA, t1=NA, c1=NA, n=NA)
   }
   atc$iter<-iter
   atc$enh0<-r$enh0[1]
   atc$EnHO<-EnHO
   atc$tai<-ta.lim[1]
   atc$tas<-ta.lim[2]
   atc$ti<-t1.lim[1]
   atc$ts<-t1.lim[2]
   atc$ci<-c1.lim[1]
   atc$cs<-c1.lim[2]
   atc$cote<-cote
   if (iter==1) {
      atcs<-atc
   } else {
      atcs<-rbind(atcs, atc)
   }
}
atcs$i<-1:dim(atcs)[1]
a<-atcs[!is.na(atcs$n),]

p<-t(apply(a,1,fct,ceps=ceps,alphaeps=alphaeps,nbmaxiter=nbmaxiter,
       dist=dist))
p<-as.data.frame(p)
names(p)<-c("cL","cU","alphac","betac","rho0","rho1","cb1","cb")
p$i<-a$i

res<-merge(atcs, p, by="i", all=T)
res$pap<-round(pnorm(res$c1),4)   ## stopping prob of stage 1 ##
res$ta<-res$n/rate
res$EnHO<-(res$ta-pmax(0, res$ta-res$t1)*res$pap)*rate
res$diffc<-res$cU-res$cL
res$c<-round(res$cL+(res$cU-res$cL)/2,4)
res$diffc<-ifelse(res$diffc<=ceps, 1, 0)
res<-res[order(res$enh0),]
des<-round(res[1,],4)

param<-data.frame(shape=shape,S0=S0,hr=hr,alpha=alpha,beta=beta,
                  rate=rate, x0=x0, tf=tf)
Two_stage<- data.frame(n1= ceiling(des$t1*param$rate), c1=des$c1,
            n=ceiling(des$ta*param$rate), c=des$c, t1=des$t1,
            MTSL=des$ta+param$tf, ES=des$EnHO, PS=des$pap)
DESIGN<-list(param=param,Single_stage=Single_stage,Two_stage=Two_stage)
return(DESIGN)
}
Optimal.KJ.MOSLRT(shape=0.5,S0=0.3,x0=1,hr=0.65,tf=1,rate=10,alpha=0.05,beta=0.2,
          dist="WB")
```

Bibliography

Anderson PK, Borgan O, Gill RD, Keiding N. *Statistical Methods Based on Counting Processes*. Springer-Verlag Inc. New York, 1993.

Belin L, Rycke YD, Broet P. A two-stage design for phase II trials with time-to-event endpoint using restricted follow-up. *Contemporary Clinical Trials Communications*, 2017;8:127-134.

Bonetti A, Zaninelli M, Leone R, et al. The use of the growth modulation index to evaluate the activity of oxaliplatin added to a 5FU-based regimen in fluoropyrimidine-resistant colorectal cancer. Proceeding of the AACR, NCI, EORTC International Conference Molecular Targets and Cancer Therapeutics, November 16-19, 1999, Washington, DC. Abstract 222.

Breslow NE. Analysis of survival data under the proportional hazards model. *Int. Statist. Rev.*, 1975; **43**:44-58.

Breslow NE, Crowley J. A large sample study of the life table and product limit estimates under random censorship. *The Annals of Statistics*, 1974; **2**:437-453.

Brown BW, et al., STPlan version 4.5; The University of Texas M.D. Anderson Cancer Center; 2010.

Case LD, Morgan TM. Design of phase II cancer trials evaluating survival probabilities. *BMC Medical Research Methodology*, 2003; **3**:1-12.

Chu C, Liu S, Rong A. Study design of single-arm phase II immunotherapy trials with long-term survivors and random delayed treatment effect. *Pharmaceutical Statistics*, 2020, **19**:358-369.

Cochran WG. The χ^2 test of goodness of fit. *Annals of Mathematical Statistics* 1952;**23**:315-345.

Collett D. *Modeling Survival Data in Medical Research*. 2nd Edition, Chapman and Hall, London, 2003.

Cook TD, DeMets D. *Introduction to Statistical Methods for Clinical Trials*. CRC Press, New York, 2007.

Corbière F, Joly P. A SAS macro for parametric and semiparametric mixture cure models. *Computer Methods and Programs in Biomedicine* 2007; **85**:173-180.

Cotterill A, Whitehead J. Bayesian methods for setting sample sizes and choosing allocation ratios in phase II clinical trials with time-to-event endpoints. *Statistics in Medicine* 2015; **34**:1889-1903.

Dixon DO, Simon R. Sample size considerations for studies comparing survival curves using historical controls. *Journal of Clinical Epidemiology* 1988; **41**:1209-1213.

Emens LA, Butterfield LH, Hodi Jr. FS, Marincola FM, Kaufman HL. Cancer immunotherapy trials: leading a paradigm shift in drug development. *Journal for ImmunoTherapy of Cancer* 2016; **4**:42.

Ewell M, Ibrahim JG. The large sample distribution of the weighted log rank statistic under general local alternatives. *Lifetime Data Analysis* 1997; **3**:5-12.

Farewell VT. The use of mixture models for the analysis of survival data with long-term survivors. *Biometrics* 1982; **38**:1041-1046.

Findlay B, Tonkin K, Crump M, Norris B, Trudeau M, Blackstein M, et al. A dose escalation trial of adjuvant cyclophosphamide and epirubicin in combination with 5-fluorouracil using G-CSF support for premenopausal women with breast cancer involving four or more positive nodes. *Annals of Oncology* 2007; **18**:1646-1651.

Finkelstein DM, Muzikansky A, Schoenfeld DA. Comparing survival of a sample to that of a standard population, *J Natl Cancer Inst,* 2003; **95**:1434-1439.

Fleming TR, Harrington DP. *Counting Processes and Survival Analysis.* Wiley, New York, 1991.

Fleming TR, Rothmann MD, Lu HL. Issues in using progression-free survival when evaluating oncology products. *Journal of Clinical Oncology* 2009; **27**:2874-2880.

George SL, Desu MM. Planning the size and duration of a trial studying the time to some critical event. *Journal of Chronic Disease* 1973; **27**:15-24.

Guyot P, Ades AE, Ouwens M, Welton NJ. Enhanced secondary analysis of survival data: reconstructing the data from published Kaplan-Meier survival curves. *BMC Medical Research Methodology* 2012; **12**:9.

Halpern J, Brown BW. Designing clinical trials with arbitrary specification of survival functions and for the log rank or generalized Wilcoxon test. *Controlled Clinical Trials* 1987; **8**:177-189.

Hauck WW, Hyslop , Anderson S. Generalized treatment effects for clinical trials. *Statistics in Medicine* 2000;**19**:887-99.

Huang B, Thomas N. Optimal designs with interim analyses for randomized studies with long-term time-specific endpoints. *Statistics in Biopharmaceutical Research*, 2014; **6**:175-184.

Huang B, Talukder E, Thomas N. Optimal two-stage phase II designs with long-term endpoints. *Statistics in Biopharmaceutical Research*, 2010; **2**:51-61.

Ibrahim JG, Chen MH, Sinha D. *Bayesian Survival Analysis*, Springer-Verlag Inc. New York, 2001.

Jennison C, Turnbull BW. *Group Sequential Methods with Applications to Clinical Trials*. Chapman and Hall, New York, 2000.

Julious SA. *Sample Sizes for Clinical Trials*. CRC Press, Florida, 2010.

Jung SH. *Randomized Phase II Cancer Clinical Trials*. CRC Press, Florida, 2013.

Jung SH, Chow SC. On sample size calculation for comparing survival curves under general hypothesis testing. *Journal of Biopharmaceutical Statistics* 2012; **22**:485-495.

Jung SH, Kim C, Chow SC. Sample size calculation for the log-rank tests for multi-arm trials with a control. *Journal of the Korean Statistical Society* 2008; **37**:11-22.

Kalbfleisch JD, Prentice RL. *The Statistical Analysis of Failure Time Data*. New York: Wiley & Sons; 1980.

Kalbfleisch JD, Prentice RL. *The Statistical Analysis of Failure Time Data*. 2nd Edition. Wiley, New York. 2002.

Kantoff PW, Higano CS, Shore ND, Berger ER, Small EJ, et al. Sipuleucel-T immunotherapy for castration-resistant prostate cancer. *The New England Journal of Medicine*, 2010; 363:411-422.

Kaplan EL, Meier P. Nonparametric estimation from incomplete observations. *Journal of the American Statistical Association* 1958; **53**:457-481.

Kooperberg C, Stone CJ. Logspline density estimation for censored data. *Journal of Computational and Graphical Statistics* 1992; **1**:301-328.

Kovalchik S, Mietlowski W. Statistical methods for a phase II oncology trial with a growth modulation index (GMI) endpoint *Contemporary Clinical Trials* 2011;**32**:99-107.

Kuk AYC, Chen CH. A mixture model combining logistic regression with proportional hazards regression. *Biometrika* 1992; **79**:531-541.

Kwak M, Jung SH, Phase II clinical trials with time-to-event endpoints: Optimal two-stage designs with one-sample log-rank test, *Statistics in Medicine* 2014;**33**:2004-2016.

Lachin JM. Introduction to sample size determination and power analysis for clinical trials. *Controlled Clinical Trials* 1981; **2**:93-114.

Lachin JM, Foulkes MA. Evaluation of sample size and power for analyses of survival with allowance for nonuniform patient entry, losses to follow-up, noncompliance, and stratification. *Biometrics* 1986; **42**:507-519.

Lakatos E. Sample sizes based on the log-rank statistic in complex clinical trials. *Biometrics* 1988; **44**:229-241.

Lakatos E. Designing complex group sequential survival trials. *Statistics in Medicine* 2002; **21**:1969-1989.

Lawless JF. *Statistical Methods for Lifetime Data*, New York: John Wiley and Sons, 1982.

Lee JW, Sather HN. Group sequential methods for comparison of cure rates in clinical trials. *Biometrics* 1995; **51**:756-763.

Lee FT, Wang JW. *Statistical Methods for Survival Data Analysis*. 3rd Edition. Wiley, New York, 2003.

Lee L. Multivariate distribution having Weibull properties. *Journal of Multivariate Analysis* 1979;**9**:267-277.

Lehmann EL. *Testing Statistical Hypotheses*. New York: Wiley and Sons, 1959.

Levine MN, Pritchard KI, Bramwell VHC, Shepherd LE, Tu D, Paul N. Randomized trial of intensive cyclophosphamide, epirubicin, and fluorouracil chemotherapy compared with cyclophosphamide, methotrexate, and fluorouracil in premenopausal women with node- positive breast cancer. *Journal of Clinical Oncology* 1998; **16**:2651-2658.

Lin DY, Shen L, Ying Z, Breslow NE. Group sequential designs for monitoring survival probabilities. *Biometrics*, 1996; **52**:1033-1042.

Lin DY, Yao Q, Ying Z. A general theory on stochastic curtailment for censored survival data. *Journal of the American Statistics Association* 1999; **94**:510-521.

Lin R, Coleman RL, Yuan Y. TOP: Time-to-event Bayesian optimal phase II trial design for cancer immunotherapy. *J Natl Cancer Inst* 2020; **112**:38-45.

Lu JC, Bhattacharyya GK. Inference procedures for a bivariate exponential model of Gumbel. *Journal of Statistical Planning and Inference* 1991;**27**:383-396.

Mick R, Crowley JJ, Carroll RJ. Phase II clinical trial design for noncytotoxic anticancer agents for which time to disease progression is the primary endpoint. *Control Clinical Trials* 2000;**21**:343-359.

Nagashima K, Noma H, Sato Y, Gosho M. Sample size calculations for single-arm survival studies using transformations of the Kaplan-Meier estimator *Pharmaceutical Statistics*, 2020; https://doi.org/10.1002/pst.2090.

O'Brien PC. Comparing two samples: extensions of the t, rank-sum, and log-rank tests. *Journal of the American Statistical Association* 1988;**83**:52-61.

Owen WJ. A power analysis of tests for paired lifetime data. *Life Time Data Analysis* 2005;**11**:233-243.

Owzar K, Jung SH. Designing phase II studies in cancer with time-to-event endpoints, *Clinical Trials*, 2008; **5**:209-221.

Peng Y, Dear KBG, Denham JW. A generalized *F* mixture model for cure rate estimation. *Statistics in Medicine* 1998; **17**:813-830.

Peng Y, Dear KBG. A nonparametric mixture model for cure rate estimation. *Biometrics* 2000; **56**:237-243.

Rubinstein LV, Gail MH, Santner TJ. Planning the duration of a comparative clinical trial with loss to follow-up and a period of continued observation, *Journal of Chronic Diseases* 1981; **34**:469-479.

Schmidt R, Kwiecien R, Faldum A, Berthold F, Hero B, Ligges S. Sample size calculation for the one-sample log-rank test. *Statistics in Medicine*, 2015;15:1031-1040.

Schoenfeld DA. The asymptotic properties of nonparametric tests for comparing survival distributions. *Biometrika* 1981; **68**:316-319.

Schoenfeld DA. Sample-size formula for the proportional-hazards regression model, *Biometrics* 1983; **39**:499-503.

Schoenfeld DA, Ritcher JR. Nomograms for calculating the number of patients needed for a clinical trial with survival as an endpoint. *Biometrics* 1982; **38**:163-170.

Seymour L. et al. The design of phase II clinical trials testing cancer therapeutics: consensus recommendations from the clinical trial design task force of the National Cancer Institute investigational drug steering committee. *Clinical Cancer Research* 2010; **16**:1710-1718.

Shan G. Two-stage optimal designs based on exact variance for a single-arm trial with survival endpoints. *Journal of Biopharmaceutical Statistics* 2020; **30**:797-805.

Shan G, Zhang H. Two-stage optimal designs with survival endpoint when the follow-up time is restricted. *BMC Medical Research Methodology* 2019; **19**(1) doi.org/10.1186/s12874-019-0696-x

Shi H, Yin G. Bayesian enhancement two-stage design for single-arm phase II clinical trials with binary and time-to-event endpoints. *Biometrics* 2018; **74**:1055-1064.

Simon R. Genomic alteration-driven clinical trial designs in oncology. *Annals of Internal Medicine* 2016;**165**:270-278.

Slud EV. Sequential linear rank tests for two-sample censored survival data. *Annals of Statistics* 1984; **12**:551-571.

Sprott DA. Normal likelihoods and relation to a large sample theory of estimation. *Biometrika* 1973; **60**:457-465.

Sun XQ, Peng P, Tu DS. Phase II cancer clinical trial with a one-sample log-rank test and its corrections based on the Edgeworth expansion. *Contemporary Clinical Trials*, 2011; **32**:108-113.

Sy JP, Taylor JMG. Estimation in a Cox proportional hazards cure model. *Biometrics* 2000; **56**:227-236.

Tarone RE, Ware J. On distribution-free tests for equality of survival distributions. *Biometrika*, 1977; **64**:156-160.

Texier M, Rotolo F, Ducreux M, Bouche O, Pignon JP, Michiels S. Evaluation of treatment effect with paired failure times in a single-arm phase II trial in oncology. *Computational and Mathematical Methods in Medicine* 2018; https://doi.org/10.1155/2018/1672176

Thall PF, Wooten LH, Tannir NH. Monitoring event times in early phase clinical trials: some practical issues, *Clinical Trials*, 2005; **2**:467-478.

Von Hoff DD. There are no bad anticancer agents, only bad clinical trial designs–Twenty-first Richard and Hinda Rosenthal Foundation Award Lecture. *Clinical Cancer Research* 1998;**4**:1079-1086.

Von Hoff DD, Stephenson JJ. Jr, Rosen P, et al. Pilot study using molecular profiling of patients' tumors to find potential targets and select treatments for their refractory cancers. *Journal of Clinical Oncology* 2010; **28**:4877-4883.

Wang S, Zhang J, Lu W. Sample size calculation for the proportional hazards cure model. *Statistics in Medicine* 2012; **31**:3959-3971.

Whitehead J. One-stage and two-stage designs for phase II clinical trials with survival endpoints. *Statistics in Medicine* 2014; **33**:3830-3843.

Wu J. A new one-sample log-rank test. *Journal of Biometrics and Biostatistics*, 2014, **5:4**:1-5.

Wu J. Sample size calculation for the one-sample log-rank test. *Pharmaceutical Statistics*, 2015, 14:26-33.

Wu J. *Statistical Methods for Survival Trial Design: With Applications to Cancer Clinical Trials Using R*. CRC Press, Florida, 2018.

Wu J, Chen L, Wei J, Weiss H, Chauhan A, Two-stage single-arm phase II survival trial design. *Pharmaceutical Statistics*, 2020, **19**:214-229.

Wu J, Chen L, Wei J, Weiss H, Miller R, Villano JL, Phase II trial design with growth modulation index as the primary endpoint. *Pharmaceutical Statistics*, 2019, **18**:212-222.

Wu J, Xiong X. Group sequential design for randomized phase III trials under the Weibull model, *Journal of Biopharmaceutical Statistics* 2015; **25**:1190-1205.

Wu J, Xiong X. Single-arm phase II group sequential trial design with survival endpoint at a fixed time point. *Statistics in Biopharmaceutical Research*, 2014;6:289-301.

Xiong X, Wu J. A novel sample size formula for the weighted log-rank test under the proportional hazards cure model. *Pharmaceutical Statistics* 2017; 16:87-94.

Xu Z, Zhen B, Park Y, Zhu B. Designing therapeutic cancer vaccine trials with delayed treatment effect. *Statistics in Medicine* 2016; **36**:592-605.

Yateman NA, Skene AM. Sample sizes for proportional hazards survival studies with arbitrary patient entry and loss to follow-up distributions. *Statistics in Medicine* 1992; **11**:1103-1113.

Yu B, Tiwari RC, Cronin KA, Feuer EJ. Cure fraction estimation from the mixture cure models for grouped survival data. *Statistics in Medicine* 2004; **23**:1733-1747.

Yuan Y, Guo B, Munsell M, Lu K, Jazaeri A. MIDAS: A practical Bayesian design for platform trials with molecularly targeted agents. *Statistics in Medicine* 2016; **35**:3892-3906.

Zhang D, Quan H. Power and sample size calculation for log-rank test with a time lag in treatment effect. *Statistics in Medicine* 2009; **28**:864-879.

Zhao L, Taylor JMG, Schuetze SM. Bayesian decision theoretic two-stage design in phase II clinical trials with survival endpoint. *Statistics in Medicine* 2012; **31**:1804-1820.

Zhou H, Chen C, Sun L, Yuan Y. Bayesian optimal phase II clinical trial design with time-to-event endpoint. *Pharmaceutical Statistics*, 2020; **19**:776-786.

Index

Milton Keynes UK
Ingram Content Group UK Ltd.
UKHW040109071024
449327UK00019B/930